Total-Reflection X-Ray
Fluorescence Analysis

CHEMICAL ANALYSIS

A SERIES OF MONOGRAPHS ON
ANALYTICAL CHEMISTRY AND ITS APPLICATIONS

Editor
J. D. WINEFORDNER

VOLUME 140

A WILEY-INTERSCIENCE PUBLICATION

JOHN WILEY & SONS, INC.

New York / Chichester / Brisbane / Toronto / Singapore / Weinheim

Total-Reflection X-Ray Fluorescence Analysis

REINHOLD KLOCKENKÄMPER

Institut für Spektrochemie und Angewandte Spektroskopie
Dortmund, Germany

A WILEY-INTERSCIENCE PUBLICATION

JOHN WILEY & SONS, INC.

New York / Chichester / Brisbane / Toronto / Singapore / Weinheim

This text is printed on acid-free paper.

Copyright © 1997 by John Wiley & Sons, Inc.

All rights reserved. Published simultaneously in Canada.

Library of Congress Cataloging in Publication Data:

Klockenkämper, R.
    Total-reflection X-ray fluorescence analysis / R. Klockenkämper.
        p.   cm.—(Chemical analysis; v. 140)
    "A Wiley-Interscience publication."
    Includes bibliographical references and index.
    ISBN 0-471-30524-3 (cloth: alk. paper)
    1. X-ray spectroscopy. 2. Fluorescence spectroscopy. I. Title.
    II. Series.
    QD96.X2K58 1996
    543'.08586—dc20   96-20190

Printed in the United States of America

10   9   8   7   6   5   4   3   2   1

CONTENTS

FOREWORD

Within the past decade, total-reflection X-ray fluorescence (TXRF) has achieved remarkable success in element analysis. It is a universal and economic multielement method suitable for micro- and trace analyses. In contrast to conventional XRF, the energy-dispersive variant TXRF is a microtechnique similar to ET-AAS. With restriction to small sample volumes, a significant improvement in detection power has been realized at the same time. Moreover, matrix effects do not occur making possible simple and reliable quantification. Compared to other promising developments, for example, laser spectroscopy, TXRF has already been well tried in several laboratories and factories. In various intercomparisons, TXRF has competed quite well with firmly established methods such as ET-AAS, ICP-OES, and INAA. The availability of commercial computer-controlled instruments will further promote the spread of this method.

TXRF has also been successfully introduced in the analysis of flat sample surfaces and stratified near-surface layers. As a nondestructive method, it is especially suitable for quality control of wafers in the semiconductor industry. In addition to composition, the nanometer thickness and density of thin layers can be determined by variable glancing-angle measurements. Direct density measurements are a special and unique feature of TXRF.

The author has been a successful and well-established expert in the field of TXRF for many years. He describes the principles and fundamentals of TXRF, the performance of TXRF analyses, and possible applications. Moreover, he critically evaluates the latest developments of this fairly new method, which is not yet sufficiently known in applied analysis. In my opinion, this book is important in that it shows TXRF to be at the leading edge of analytical atomic spectrometry.

This well-written first monograph on TXRF provides valuable assistance to present and future users of this highly powerful analytical tool. TXRF enables economical and reliable determination of very low absolute amounts of elements in the ng and pg range in many inorganic and organic matrices. Therefore many disciplines such as the geo-, bio-, material, and environmental sciences, as well as medicine, toxicology, forensics, and archaeometry, will profit when applying TXRF to analytical problems concerning ultramicro- and trace analyses of elements.

Dortmund, Germany
February 1996

PROF. GÜNTHER TÖLG

Director, Retired
Institute of Spectrochemistry and
Applied Spectroscopy

ACKNOWLEDGMENTS

The author is indebted to A. von Bohlen and M. Becker for valuable assistance, for careful drawing of the figures, for recording and evaluating spectra, for the photographs, and for typing the manuscript. Further scientific and technical assistance was given by the Institut für Spektrochemie und Angewandte Spektroskopie (ISAS). The Institute in Dortmund is supported by the Bundesministerium für Bildung, Wissenschaft, Forschung und Technologie (BMBF) of Germany and by the Ministerium für Wissenschaft und Forschung (MWF) of North Rhine–Westphalia.

CHEMICAL ANALYSIS

A SERIES OF MONOGRAPHS ON
ANALYTICAL CHEMISTRY AND ITS APPLICATIONS

J. D. Winefordner, *Series Editor*

xiii

Total-Reflection X-Ray
Fluorescence Analysis

CHAPTER

1

FUNDAMENTALS OF X-RAY FLUORESCENCE

X-ray fluorescence (XRF) is based on the irradiation of a sample by a primary X-ray beam. The individual atoms hereby excited emit secondary X-rays that can be detected and recorded in a spectrum. The spectral lines or peaks of this spectrum are characteristic of the individual atoms i.e., of the respective elements present in the sample so that, by an appropriate interpretation of the spectrum, the sample can be analyzed.

XRF analysis reaches near-surface layers of only about 100 μm thickness but generally is performed without any consumption of the sample. The method is fast and can be applied universally to a great variety of samples. Solids can be analyzed directly with no or only little sample preparation. Apart from the light elements, all elements with atomic numbers greater than 11 (possibly greater than 6) can be detected. The method is sensitive down to the microgram-per-gram level, and the results are precise and also accurate if matrix effects can be corrected.

For these reasons, XRF has become a well-known method of spectrochemical analysis. It plays an important role in the industrial production of materials, in prospecting mineral resources, and recently in environmental monitoring. The number of spectrometers in use are estimated to be about 15,000 worldwide. Of these, 80% are operating in the wavelength-dispersive mode with analyzing crystals; only 20% operate in the energy-dispersive mode, mainly with Si(Li) detectors. At present, however, energy-dispersive spectrometers are four times more frequently built than wavelength-dispersive instruments due to the advantage the former provide in fast registration of the total spectrum.

1.1. HISTORY AND RELATIONS OF XRF AND TXRF

The foundations of spectrochemical analysis were laid by R.W. Bunsen, a chemist, and G.R. Kirchhoff, a physicist. In 1859, they vaporized a salt in a flame and determined some alkaline and alkaline-earth metals by means of an optical spectroscope. Today, optical atomic spectroscopy has developed a variety of new analytical techniques with high efficiency such as atomic absorption spectroscopy (AAS) with flames (FAAS) or electrothermal furnaces

1

(ET-AAS) and the inductively coupled plasma technique (ICP) combined with optical emission or mass spectrometry (ICP-OES and ICP-MS). These techniques do entail some consumption of the sample, but they are highly suitable for ultratrace analyses of solutions.

In 1895, three and a half decades after the discovery of Bunsen and Kirchhoff, W.C. Röntgen detected a still unknown radiation which he called X-rays. Then, in 1913, Moseley established the basis of X-ray fluorescence analysis by his well-known law [1]. It relates the wavelength of the "characteristic" X-rays to the atomic number of the elements causing this radiation. In the mid-1940s, the first X-ray spectrometers became available. In the decades since, XRF has developed into a powerful method of spectrochemical analysis, as mentioned above. However, classical XRF is not suitable for ultratrace analyses and it is notorious for producing matrix effects that may lead to systematic errors. Extensive efforts have been made to overcome these drawbacks, e.g., by matrix separation, thin-film formation, and mathematical corrections. Nevertheless, the new techniques of optical atomic spectrometry have surpassed conventional XRF in many respects.

An important advance, however, was achieved by the ingenious idea of using total reflection for the excitation of XRF. Earlier, the phenomenon of total reflection for X-rays had been discovered in 1923 by Compton [2]. He found that the reflectivity of a flat target strongly increased below a critical angle of only about 0.1°. But in 1971 Yoneda and Horiuchi [3] first took advantage of this effect for XRF. They proposed the analysis of a small amount of material applied on a flat totally reflecting support. This technique was subsequently developed [4–7] and called total-reflection X-ray fluorescence (TXRF).

As is illustrated in Figure 1-1, TXRF is a variation of energy-dispersive XRF with, however, one significant difference. In contrast to XRF, where the primary beam strikes the sample at an angle of about 40°, TXRF uses a glancing angle of less than 0.1°. Owing to this grazing incidence, the primary beam shaped like a strip of paper is totally reflected.

Today, TXRF is primarily used for chemical *micro-* and *trace* analyses. For this purpose, small quantities mostly of solutions or suspensions are placed on optical flats, e.g., quartz glass, serving as sample supports. After evaporation, the residue is excited to fluorescence under the fixed small glancing angle and the characteristic radiation is recorded by a Si(Li) detector as an energy-dispersive spectrum. It is the *high reflectivity* of the sample support that nearly eliminates the spectral background of the support and lowers the detection limits from 10^{-7} to 10^{-12}g. Although this mode of operation does not permit the entirely nondestructive investigation of bulk material, it offers new challenging possibilities in ultramicro- and trace analyses. Besides its high detection power, simplified quantification is made possible by internal

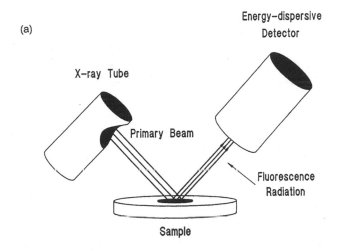

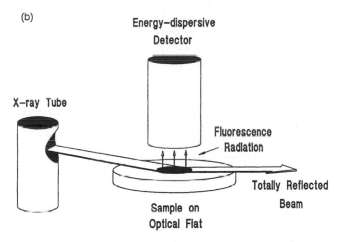

Figure 1-1. Instrumental arrangement for (a) conventional XRF and (b) TXRF. Comparison shows a difference in the geometric grouping of excitation and detection units.

standardization. This is because matrix effects cannot build up within the minute residues or thin layers of a sample.

A new field of application has been opened in the 1980s by *surface* and *near-surface layer* analyses. In 1983, an angular dependence of X-ray fluorescence intensities in the range of total reflection was first observed by Becker et al. [8]. This effect was used in the following years to investigate surface impurities, thin near-surface layers, and even molecules adsorbed on flat surfaces. Such examinations are especially applicable for cleaned and/or layered wafers representing the basic material for the semiconductor industry. The flat samples are examined either with respect to contaminations of the surface or with respect to the setup of near-surface layers. But this mode of analysis needs fluorescence intensities to be recorded not only at one fixed angle but at various angles around the critical angle of total reflection. From these angle-dependent intensity profiles, the composition, thickness, and even density of top layers can be ascertained. It is the *low penetration depth* of the primary beam at total reflection that enables this in-depth examination of ultrathin layers in the range of 1–500 nm. The method is nondestructive and needs no vacuum—at least no ultrahigh vacuum (UHV).

In spite of the similarities in instrumentation such as the X-ray source, the energy-dispersive detector, and pulse-processing electronics, the use of TXRF differs fundamentally from classical XRF. With respect to sample preparation and performance of analysis, it has a lot in common with AAS or ICP for trace element analysis and it is similar to X-ray photoelectron spectroscopy (XPS), Rutherford backscattering spectroscopy (RBS), and secondary ion mass spectrometry (SIMS) for surface and near-surface layer analysis. In fact, TXRF is able to compete, often favorably, with these well-established methods.

The main reason for this progress is the special geometric arrangement leading to total reflection of the primary beam. Accordingly, the totally reflected beam interferes with the incident primary beam and leads to standing waves above surfaces and also within near-surface layers. The unique role of TXRF is based on the formation of such standing waves and particular details can only be understood with regard to these standing waves.

The arrangement of grazing incidence is not restricted to XRF measurements. It can also be exploited for X-ray reflection (XRR) and X-ray diffraction (XRD). As early as 1931 Kiessig investigated the reflection of thin layers deposited on a thick substrate [9], and in 1940 DuMond and Youtz observed Bragg reflection of periodic multilayers [10]. It was not until the late 1970s that XRD at grazing incidence was developed. This monograph mainly deals with the technique of TXRF and excludes that of XRD. But reports of XRR experiments are included when needed for a better understanding or even for complementary results. The usual TXRF instrumentation can simply be extended for such experiments.

In recent years, the considerable advantages of TXRF have made it a well-established method of spectrochemical analysis. A great variety of applications have promoted a growing interest. At least four different instruments are commercially available today (EXTRA II of Seifert, Model 3726 of Rigaku, TREX 600 of Technos, and TXRF 8010 of Atomika). The number of instruments in use is estimated to be about 300 worldwide. Five workshops have taken place: three in Germany, in Geesthacht, 1986 and 1992, and in

Dortmund, 1988; one in Vienna, 1990; and one in Tsukuba near Tokyo, 1994. The papers presented were subsequently published in special issues of scientific journals [11–14].

1.2. NATURE AND PRODUCTION OF X-RAYS

As already noted, X-rays are part of the electromagnetic radiation, discovered by Röntgen in 1895. They can be described partly in a corpuscle picture and partly in a wave picture. In the corpuscle picture, radiation is transmitted by quanta of energy called *photons*. A photon is a corpuscle that carries an elementary energy unit E but has no rest mass. *In vacuo*, all photons travel at the velocity of light c on straight lines that can be regarded as X-ray beams. In the wave picture, X-rays propagate as waves showing crests and troughs of the field strength. They follow each other with a frequency v and at a distance λ called the wavelength and are always orthogonal to the direction of the respective beam.

X-ray photons have energies in the kiloelectronvolt range (0.1–100 keV). Energy and frequency are directly proportional to each other, and energy and wavelength are inversely so, according to

$$E = hv = \frac{hc}{\lambda} \qquad (1\text{-}1)$$

where h is the Planck's constant ($h = 4.1357 \times 10^{-18}$ keV·s; $c = 2.9979 \times 10^{8}$ m/s). The conversion of energy and wavelength can be made by the relationship

$$E[\text{keV}] = \frac{1.2397}{\lambda[\text{nm}]} \qquad (1\text{-}2)$$

X-rays are originally produced by the bombardment of matter with accelerated electrons. Usually, such a *primary* radiation is produced by an X-ray tube of the Coolidge type, as shown in Figure 1-2. It consists of a vacuum-sealed tube with a metal–glass cylinder. A tungsten filament serves as the hot cathode, and a pure-metal target such as chromium, copper, molybdenum, or tungsten serves as the anode. Electrons are emitted from the heated filament and accelerated by an applied high voltage in the direction of the anode. The high-energy bombardment of the target produces X-rays that emerge from a thin exit window as an intense X-ray beam. Mostly, a 0.2–1 mm thick beryllium window is used.

The X-ray tube is supplied by a stabilized high-voltage generator. High voltage and current applied to the tube determine the intensity of the X-ray

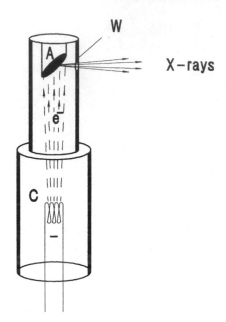

Figure 1-2. X-ray tube of the Coolidge type used as an X-ray photon source: C = tungsten filament used as the cathode; A = metal block with a slant plane used as the anode; W = thin exit window.

beam. The voltage can usually be chosen between 10 and 60 or even 100 kV, the current between 10 and 50 mA, so that an electric power of several kilowatts can be supplied. However, only about 0.1% of the electric input power is converted into radiation and most of it is dissipated as heat. For that reason, such X-ray tubes have to be cooled intensively by water. A flow rate of 3 to 5 L/min is commonly needed.

The primary X-ray beam is normally used to irradiate a sample for analysis. By this primary irradiation, the atoms in the sample are generally excited to produce *secondary* X-rays by themselves. This effect is called X-ray fluorescence. The secondary radiation can be recorded as a color pattern of the sample as its *chromatic* composition changes with the *element* composition. The spectral pattern recorded by means of an X-ray detector constitutes the basis of XRF analysis.

X-ray spectra generally show the intensity of radiation or rather the number of its photons in relation to the wavelength of radiation or the energy of its photons. Normally, X-ray spectra consist of two different parts, the line spectrum and the continuous spectrum.

1.2.1. The Line Spectrum

A line spectrum will be produced if a target is irradiated with X-ray photons, as just mentioned, or is bombarded with electrons (or ions) of a sufficient energy. The energy must exceed the binding energy of a bound inner electron of the target atoms, which therefore is called the *critical excitation energy*. The ensuing effects can be described best by Niels Bohr's atomic model, which supposes Z electrons revolving around a nucleus in different orbitals or shells and subshells, where Z is the atomic number of the respective element.

Owing to the high-energy impact, an inner electron can be ejected from the atom so that a vacancy is created within the relevant inner electron shell. The atom with the vacancy is in an unstable state of higher energy and needs to regain its stable ground state by two different processes. In both processes an outer bound electron fills the vacancy and the atom instantly emits either an X-ray photon, which is the basic process of XRF, or what is called an Auger electron. The energy of the X-ray photon must be equal to the difference of the previous and the subsequent energy state of the atom:

$$E_{\text{photon}} = E_{\text{previous}} - E_{\text{subsequent}} \tag{1-3}$$

The newly created vacancy in the outer shell can be filled in turn by an electron still farther out, and another X-ray photon can be emitted. These processes will follow each other successively and a series of photons will be emitted until a free electron ultimately replaces an outermost valence electron so that the atom has finally returned to the ground state.

Since the energy states of atomic electrons are quantized and characteristic of all atoms of an element, the X-ray photons emitted in this way have individual energies that are equal for all atoms of the same element but different for atoms of different elements. Consequently, these photons cause discrete sharp lines or peaks as intensity maxima in an X-ray spectrum that are characteristic for any single element. Conversely, any element can be identified by its characteristic lines or peaks, comparable to a fingerprint. For this reason, the line spectra are also called characteristic spectra.

Although not every electron is permitted to fill an inner vacancy, there are a lot of allowed transitions according to the selection rules of quantum theory. The most important transitions are indicated in Figure 1-3 (see, e.g., Bertin [15]). They lead to the principal lines or peaks named here in the classical notation. There are three principal series, the K-, L-, or M-series, which arise when the inner vacancy being filled is in the K-, L-, or M-shell. A series contains several peaks named K-, L-, or M-peaks, which mainly differ according to the origin of the outer electron. The most intense peak is called α; the next less intense peaks in descending order are called β, γ, η, and l.

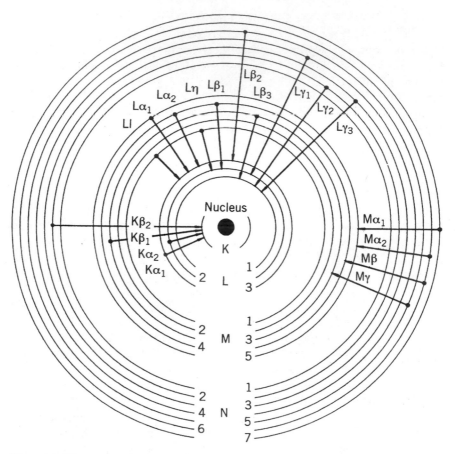

Figure 1-3. Electron transitions that are basically possible in a heavy atom and that produce the principal lines or peaks within an X-ray spectrum.

A further differentiation is made by an arabic numeral added as an index, such as α_1 and α_2 for the α-doublet. This classical notation proposed by K.M.G. Siegbahn after 1920 is not entirely systematic and indeed somewhat confusing. Meanwhile, the International Union of Pure and Applied Chemistry (IUPAC) has suggested a nomenclature that is solely based on the shell designation. Table 1-1 compares the K- and L-peaks in both notations [16].

From the preceding discussion it can be understood that the two lightest elements H and He have no X-ray peaks at all because of their lack of inner electrons. But all other elements have characteristic X-ray peaks. They appear in a spectrum with an intensity that depends on the energy of the primary

Table 1-1. K- and L X-ray Lines or Peaks in Siegbahn and IUPAC Notation

Siegbahn	IUPAC	Siegbahn	IUPAC	Siegbahn	IUPAC
$K\alpha_1$	$K\text{-}L_3$	$L\alpha_1$	$L_3\text{-}M_5$	$L\gamma_1$	$L_2\text{-}N_4$
$K\alpha_2$	$K\text{-}L_2$	$L\alpha_2$	$L_3\text{-}M_4$	$L\gamma_2$	$L_1\text{-}N_2$
$K\beta_1$	$K\text{-}M_3$	$L\beta_1$	$L_2\text{-}M_4$	$L\gamma_3$	$L_1\text{-}N_3$
$K\beta_2^{I}$	$K\text{-}N_3$	$L\beta_2$	$L_3\text{-}N_5$	$L\gamma_4$	$L_1\text{-}O_3$
$K\beta_2^{II}$	$K\text{-}N_2$	$L\beta_3$	$L_1\text{-}M_3$	$L\gamma_4'$	$L_1\text{-}O_2$
$K\beta_3$	$K\text{-}M_2$	$L\beta_4$	$L_1\text{-}M_2$	$L\gamma_5$	$L_2\text{-}N_1$
$K\beta_4^{I}$	$K\text{-}N_5$	$L\beta_5$	$L_3\text{-}O_{4,5}$	$L\gamma_6$	$L_2\text{-}O_4$
$K\beta_4^{II}$	$K\text{-}N_4$	$L\beta_6$	$L_3\text{-}N_1$	$L\gamma_8$	$L_2\text{-}O_1$
$K\beta_{4x}$	$K\text{-}N_4$	$L\beta_7$	$L_3\text{-}O_1$	$L\gamma_8'$	$L_2\text{-}N_{6,7}$
$K\beta_5^{I}$	$K\text{-}M_5$	$L\beta_8$	$L_3\text{-}N_{6,7}$	$L\eta$	$L_2\text{-}M_1$
$K\beta_5^{II}$	$K\text{-}M_4$	$L\beta_9$	$L_1\text{-}M_5$	Ll	$L_3\text{-}M_1$
		$L\beta_{10}$	$L_1\text{-}M_4$	Ls	$L_3\text{-}M_3$
		$L\beta_{15}$	$L_3\text{-}N_4$	Lt	$L_3\text{-}M_2$
		$L\beta_{17}$	$L_2\text{-}M_3$	Lu	$L_3\text{-}N_{6,7}$
				Lv	$L_2\text{-}N_{6,7}$

X-rays or electrons, on the composition of the target, and on the efficiency of the detector. In the range up to 40 keV, normally each element apart from H and He shows between 2 and about 10 intensive peaks, so that X-ray spectra in contrast to ultraviolet (UV) spectra can be regarded as fortunately poor in peak number. Figure 1-4 shows some examples of different pure elements excited at 40 keV. The lighter elements up to $Z = 25$ mostly show a $K\alpha$-doublet that is not resolved here and a $K\beta$-peak at higher energy. The heavier elements with $25 < Z < 57$ additionally have several L-peaks mostly with an α-doublet followed by a more energetic β- and γ-group. Heavy elements with $Z > 57$ are lacking in K-peaks (their exciting potential is > 40 keV) but show some M-peaks in addition to the L-peaks. In general, the most intensive K- or L-peaks are used for X-ray spectral analysis.

As mentioned earlier, the relationship of peak or photon energy and element was discovered by H.G.J. Moseley in 1913. He found that the reciprocal wavelength $1/\lambda$, i.e. the photon energy E is dependent on the atomic number Z of the elements. His well-known law can be described by

$$E = k_j(Z - \sigma_j)^2 \tag{1-4}$$

with certain constant values k_j and σ_j for particular peaks or lines j. This square law is demonstrated in Figure 1-5 by different parabolas, each representing a particular peak (see, e.g., Bertin [15]).

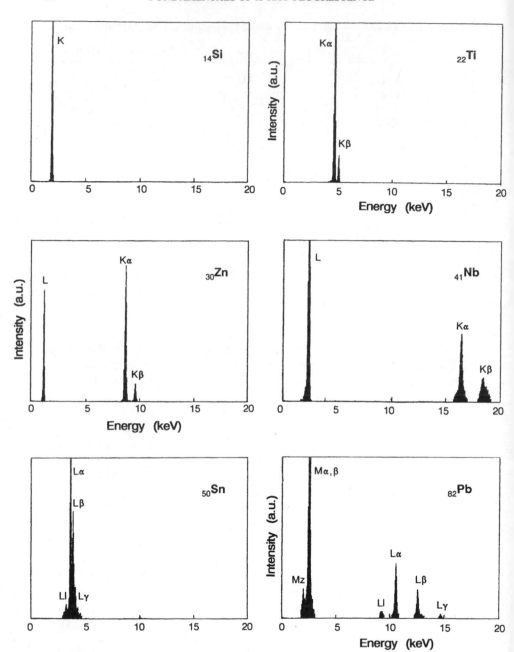

Figure 1-4. X-ray spectra with *K*-lines of silicon and titanium (above), with *L*- and *K*-lines of zinc and niobium (middle), and with *L*- and *M*-lines of tin and lead (below). The fluorescence intensity in arbitrary units is plotted against the photon energy in keV.

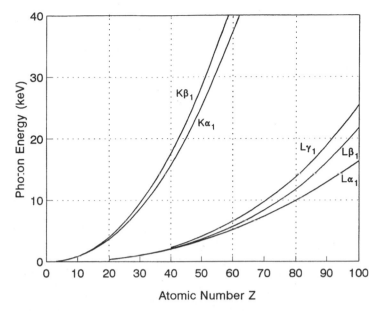

Figure 1-5. Moseley diagram of photon energies of the principal X-ray emission peaks plotted against the atomic number Z of the elements.

The quantities k_j and σ_j characterize the different peaks, although they are not exactly equal for all elements but somewhat dependent on their atomic number. The k-values for the individual $K\alpha$-peaks lie between 10 and 11 eV, for the $L\alpha$-peaks between 1.7 and 2 eV, and for the $M\alpha$-peaks at about 0.7 eV. The quantity σ can be interpreted as a shielding constant. From the point of view of an outer electron, it may be defined as that number of negative electrons by which the number of positive charges of the nucleus is reduced. For $K\alpha$-peaks it lies between 1 and 2; for $L\alpha$-peaks, between 6 and 10; and for M-peaks, at about 20.

As Moseley's law is not very stringent, the exact positions of characteristic X-ray lines are not calculated in practice by using this law but instead obtained from tables or computer memories. They normally give the energies and wavelengths of the peaks and additionally their relative intensities within the defined K-, L-, or M-series. The relative intensity of a certain peak in its series is determined by the probability of the electron transition causing this particular peak. The respective quantity is called emission rate g_j and can be calculated from quantum mechanics. In general, the relative intensities are rather similar for most elements. For the K-peaks, $K\alpha : K\beta$ is about $100:15$; for the L-peaks $Ll : L\alpha : L\eta : L\beta : L\gamma_1 : L\gamma_3$ is round $3:100:1:70:10:3$; and for the M-peaks, $M\alpha : M\beta : M\gamma$ is about $100:50:4$.

The intensity of the total K-, L-, and M-series is a function of the fluorescence yield ω. It gives the relative frequency in percent according to which an X-ray photon and not an Auger electron is emitted after excitation of an atom. The relationship can be described approximately by

$$\omega = \frac{Z^4}{A + Z^4} \qquad (1\text{-}5)$$

For the K-series, the constant A is about 9×10^5; for the L-series, 7×10^7; and for the M-series, 1×10^9.

The fluorescence yield for the K-, L-, and M-series (see, e.g., Bertin [15] or Jenkins [16]) correlates with the atomic number Z, as shown in Figure 1-6. As demonstrated there, the X-ray photon and Auger electron emission are two competing effects, the frequencies of which sum up to 100%. The Auger process predominates for lighter elements, so that X-ray spectral analysis is not very effective for those elements with atomic numbers $Z < 20$ and especially for those with $Z < 10$.

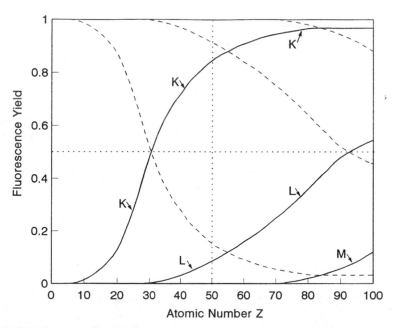

Figure 1-6. Fluorescence yield (——) and Auger electron yield (- - - - -) as a function of the atomic number Z of the emitter.

There are some exceptions to the aforementioned selection rules. First, there are emission peaks that do not correspond to permitted transitions and therefore are called "forbidden" peaks. Secondly, there are additional peaks that arise from a double ionization by a simultaneous impact of photons or electrons on two inner electrons of the atom. As the energy levels of the doubly ionized atom slightly differ from those of the singly ionized atom, somewhat different peaks occur that are called "satellite" peaks. Forbidden and satellite peaks are always weak, and satellites mainly appear in the K-spectra of lighter elements. Nevertheless, they must not be ignored in trace analysis if they are generated by a major component at energies close to small peaks of trace elements. Consequently, both forbidden and satellite peaks are also included in tables or stored in computers.

According to their energy position, the characteristic X-ray peaks are independent of the chemical bonding or state of the atoms. This advantage exists as long as only electrons from inner shells are involved in the X-ray emission process and as long as these electrons are not affected by the chemical vicinity of the atoms. In practice, this is the normal situation for the detection of high photon energies and heavier elements. However, exceptions can appear for low energies and lighter elements. If an electron from a valence or a near valence band is involved in the emission process, the respective energy level of the atom and the energy transition will be affected by the chemical state. Consequently, the characteristic peaks may be shifted for elements in different compounds. As the effect is in the range of a few electronvolts, it can be measured and used to get information on chemical bonding. However, other spectroscopic techniques are more efficient in this respect. For the usual X-ray spectrometrical practice, peak shifts are an exception but may be taken into account to avoid systematic errors.

1.2.2. The Continuous Spectrum

This kind of spectrum is defined by an intensity distributed continuously over a broad range of energy or wavelengths. For this reason, it is called a "continuous" or "white" spectrum. It is originally produced by energetic electrons or ions bombarding a target but actually not by X-ray photons themselves. However, if X-rays of a continuous spectrum are used to excite a sample, they will partly be scattered by this target and the original primary spectrum will be transformed into a somewhat modified spectrum that is likewise continuous. Consequently, a continuous spectrum is present in any case, representing an inconvenient "background" that has to be eliminated from the analytical point of view. An example of a continuous spectrum produced by an X-ray tube is given in Figure 1-7. The characteristic L-lines of the tube target are shown in addition.

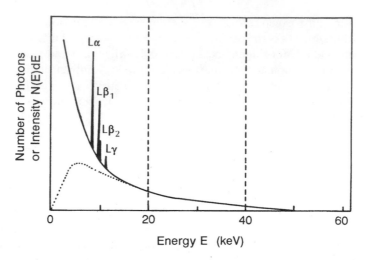

Figure 1-7. Spectrum of an X-ray tube operated at 50 kV and equipped with a thin target layer of tungsten as the anode. The spectrum is represented in the energy-dispersive mode. The *K*-lines, with an excitation energy of 69.5 keV, cannot arise; but all *L*-lines, with a required minimum of 12.1 keV, appear. The continuum of a thick solid target is illustrated by the dotted line.

The production of a spectral continuum is based on the fact that electrons penetrating into a target material are decelerated by impacts with the atomic nuclei of the target. The primary electrons lose energy in these inelastic collisions, and this energy can be emitted as X-ray photons. A single electron can lose its energy completely in a single collision or stepwise in several different collisions. Consequently, a single photon can be produced with the total electron energy or several photons can be produced with smaller parts of this energy.

In an X-ray tube operated at a voltage U_0 all electrons get the final energy E_0 according to

$$E_0 = e U_0 \tag{1-6}$$

Consequently, photons can carry away this maximum energy E_0 or lower energies down to zero. The spectrum covers the energy range between zero and the upper limit E_0, as shown in Figure 1-7. Since the deceleration of electrons causes the continuous spectrum, it is also called the *brems*-continuum (German *Bremse* = English *brake*).

The intensity distribution of the continuum can be described by

$$N(E)\,dE = k\,i\,Z(E_0/E - 1)\,dE \tag{1-7}$$

where $N(E)\,dE$ is the number of photons with energies between E and $E + dE$; k is a constant; i is the tube current; and Z is the (mean) atomic number of the target. The formula shows the intensity or number of photons inversely related to the energy E of these photons, decreasing to zero when E approximates E_0. Furthermore, it indicates that the intensity can be increased linearly by the tube current i or the voltage U_0 (by virtue of E_0) and by the atomic number Z of the target material. For that reason, high-power X-ray tubes equipped with a heavy-metal anode are frequently applied in X-ray fluorescence analysis.

Equation (1-7) can be transformed into a wavelength-dependent equation known as Kramers' formula:

$$N(\lambda)\,d\lambda = kiZ\left(\frac{\lambda}{\lambda_0} - 1\right)\frac{1}{\lambda^2}\,d\lambda \qquad (1\text{-}8)$$

where λ_0 corresponds to E_0 according to $\lambda_0 = hc/E_0$. The relationship is represented in Figure 1-8 [17], which shows a sharp short-wavelength limit at λ_0, a hump with a maximum at

$$\lambda_{max} = 2\lambda_0 \qquad (1\text{-}9)$$

and an extended long-wavelength tail.

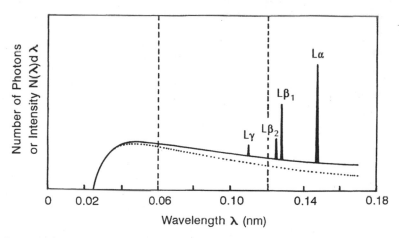

Figure 1-8. Spectrum of Figure 1-7 represented in the wavelength-dispersive mode. The X-ray tube operated at 50 kV may be equipped with a thin target layer (———) or a thick solid target (·····). (See, e.g., Tertian and Claisse [17].)

The two spectral distributions of the continuum given by equations (1-7) and (1-8) are only valid for thin target layers. For thick solid targets, they are good approximations in the high-energy or short-wavelength region (hard X-rays). For low energies or long wavelengths (soft X-rays), they are substantially modified by the self-absorption of X-rays within the target itself. These absorption losses reduce the continuous spectrum, and this effect is further increased by the absorption of X-rays in the exit window of the X-ray tube. In the end, a quite different curve results for the soft X-ray region. Figure 1-8 represents such a curve with a maximum shifted from $2\lambda_0$ to $1.8\lambda_0$. Figure 1-7 shows the corresponding continuum of the thick target with a maximum at $E = 0.1E_0$, whereas the respective distribution for a thin layer has no relative maximum at all.

1.3. ATTENUATION OF X-RAYS

Different phenomena have to be considered as forming the basis of X-ray spectrometry: the attenuation of X-rays as well as their deflection and interference. These phenomena result from the interaction between radiation and matter and can be described partly by the wave picture and partly by the corpuscle picture.

If an X-ray beam passes through matter, it will lose intensity due to different effects. According to Figure 1-9, the number N_0 of photons hitting a homogeneous sheet or layer of density ρ and thickness d is reduced to a fraction N being transmitted while the difference $\Delta N = N_0 - N$ has been lost. Generally, the attenuation of intensity is controlled by the Lambert–Beer law. This law can be written either in the differential form

$$\frac{\Delta N}{N} = -\left(\frac{\mu}{\rho}\right)\rho\,\Delta d \tag{1-10}$$

or in the integral form

$$N(d) = N_0\exp\left[-\left(\frac{\mu}{\rho}\right)\rho d\right] \tag{1-11}$$

where (μ/ρ) is called the mass-attenuation coefficient. The intensity exponentially depends on the thickness d of the layer and will be reduced to $1/e$ or nearly 37% if, for example, X-ray photons of about 20 keV pass through metal sheets of medium density and about 10–200 μm thickness.

The mass-attenuation coefficient (μ/ρ) expressed in cm^2/g is a quantity that depends on the composition (elements) of the material and the energy of the X-ray photons. Since the density ρ is incorporated, the quantity (μ/ρ) is

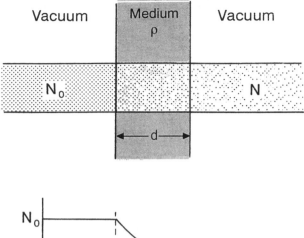

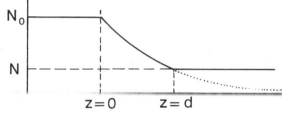

Figure 1-9. Attenuation of an X-ray beam penetrating through a homogeneous medium of density ρ and thickness d. The number of photons is reduced exponentially from N_0 to N.

independent of the state of aggregation. Values for a solid, liquid, or gas, whether it be a compound, solution, or mixture, will be equal if the composition of the material is equal. For the sake of clarity, this notation is preferred here. Some other authors use only the quantity μ instead of the product $(\mu/\rho)\rho$ for simplicity; others choose the symbol μ instead of (μ/ρ).

The mass-attenuation coefficient (μ/ρ) follows an additive law so that values of a compound, solution, or mixture can readily be calculated from values of the individual elements if the element composition is known:

$$(\mu/\rho)_{\text{total}} = \sum c_i(\mu/\rho)_i \tag{1-12}$$

where the values of c_i are the mass fractions of the different elements present in the total mixture. Of course, the individual coefficients $(\mu/\rho)_i$ are functions of the energy of the X-ray photons, so that the total value is determined only for X-ray photons of a certain energy, i.e., for a monoenergetic X-ray beam. The individual values for each element and for several energies of X-ray photons can be taken from tables (e.g., from Bertin [15] or Tertian and Claisse [17]) or can be calculated (e.g., using functions given in Williams [18]).

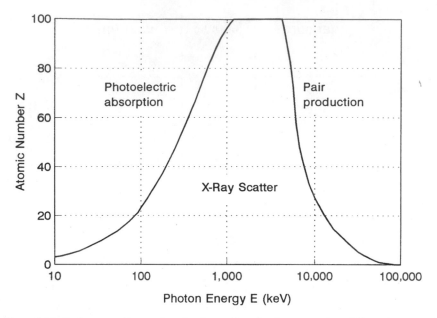

Figure 1-10. Isofrequency lines and predominant effects for the attenuation of X-rays in material of atomic number Z plotted against the photon energy E. (After Krieger and Petzold [19].)

The attentuation of X-rays is caused by the interaction of their photons with the inner or the outer electrons or even with the nuclei of atoms. It results from three competing effects called photoelectric absorption, X-ray scatter, and pair production. As indicated in Figure 1-10 [19], the photoelectric effect predominates for $E < 100$ keV and is the most important in X-ray spectroscopy. Pair production does not occur for $E < 1$ MeV, so it is insignificant for X-ray spectroscopy and will not be considered further here.

1.3.1. Photoelectric Absorption

The major component of X-ray attenuation is caused by the photoelectric effect by which an electron of an inner shell of an atom is expelled by a photon of sufficient energy. The primary photon itself is completely annihilated while a secondary photon of lower energy is emitted immediately after the electronic rearrangement. The secondary emission is called X-ray fluorescence (already described in Section 1.2.1).

Photoelectric absorption is evaluated numerically by a specific mass-absorption coefficient (τ/ρ). It can be considered as the sum of all really possible expulsions of electrons from the various atomic shells K, L, M, N,

O and P, and consequently is determined by

$$(\tau/\rho) = (\tau/\rho)_K + (\tau/\rho)_L + (\tau/\rho)_M + (\tau/\rho)_N + (\tau/\rho)_O + (\tau/\rho)_P \qquad (1\text{-}13)$$

The different additive parts can be further split up according to the corresponding subshells. All the individual coefficients approximately follow the Bragg–Pierce law:

$$(\tau/\rho)_j = k_j Z^3 / E^{8/3} \qquad (1\text{-}14)$$

with different constants k_j for the different subshells or levels j. In a double-logarithmic plot of (τ/ρ) vs. E presented in Figure 1-11, the linear segments show a negative slope of $-8/3$ and are mutually parallel. At the absorption edges, abrupt jumps of (τ/ρ) appear because further electrons of the next outer shell can be expelled if the photon energy exceeds the corresponding edge energy. For higher energies, the mass-absorption coefficient gradually falls again with the slope $-8/3$. The specific edge energies correspond to the

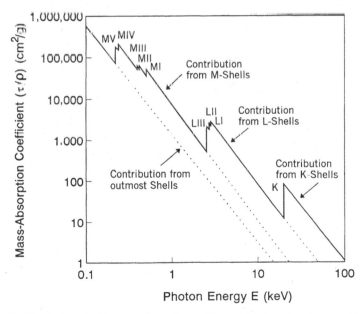

Figure 1-11. Total photoelectric mass-absorption coefficient (τ/ρ) for molybdenum vs. the photon energy E. Each discontinuity corresponds to an additional photoelectric process that occurs if the respective absorption edge K, $LI.....LIII$, or $MI....MV$ is exceeded or jumped over. (Data from Bertin [15].)

binding or ionization energies of electrons in the respective shells or subshells. They follow the Moseley law (1-4) as the emission peaks do, but with somewhat different constants. The respective emission peaks always lie at somewhat lower energies.

The jump ratio r_j at an absorption edge is defined by the quantity

$$r_j = \frac{(\tau/\rho)_{\text{high}}}{(\tau/\rho)_{\text{low}}} \tag{1-15}$$

where the subscripts "high" and "low" refer to the high- and low-energy side of an edge. The jump ratios of the K- and L-edges are represented in Figure 1-12 for various elements. From the jump ratio r_j, another useful quantity can be

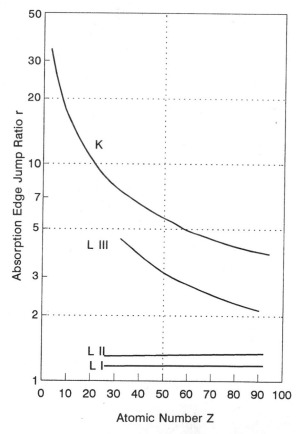

Figure 1-12. Absorption-edge jump ratio r for K-, LI, LII-, and LIII-edges plotted against the atomic number Z of the material. (After Bertin [15].)

derived called the absorption jump factor f_j. It is defined as the fraction $(\tau/\rho)_j$ of a certain shell or subshell j with respect to the total value (τ/ρ) according to

$$f_j = \frac{(\tau/\rho)_j}{\sum (\tau/\rho)_j} \qquad (1\text{-}16)$$

Jump factor f_j and jump ratio r_j are correlated according to

$$f_j = \frac{r_j - 1}{r_j} \qquad (1\text{-}17)$$

1.3.2. X-ray Scatter

The second and generally minor component of X-ray attenuation is caused by the scattering of X-ray photons. In contrast to photoelectric absorption, the primary photons do not ionize an atom but are deflected from their original direction. Two processes can be distinguished:

1. The collision of a photon with a firmly bound inner electron of an atom, leading to a change of direction of the photon without energy loss; this process is called *elastic scattering* or *Rayleigh scattering*.
2. The collision of a photon with a loosely bound outer electron or even with a free electron, leading to a change of direction *and* a loss of energy; this process is called *inelastic scattering* or *Compton scattering*.

Generally, the photons can be deflected in all directions. Rayleigh scattering can be coherent, i.e., there is a fixed relation of phases for the incident photons and the scattered photons. It occurs for scattering at crystal planes or multilayer interfaces (Sections 2.1.2 and 2.2.3). By way of contrast, Compton scattering is always incoherent.

The loss of *energy* a photon suffers in Compton scattering results from the conservation of total energy and total momentum at the collision of the photon and the electron. A photon with the energy E keeps the part E' when it is deflected by an angle ψ while the electron takes off the residual part of energy $dE = E - E'$. The fraction E'/E can be calculated according to

$$E'/E = 1/[1 + (1 - \cos\psi)E/E_e] \qquad (1\text{-}18)$$

Here E_e is the rest energy of an electron, which amounts to $E_e = m_e c^2 = 511$ keV, where m_e is the rest mass of an electron. Figure 1-13 represents the distribution of E'/E for any given direction ψ in polar coordinates. The ratio depends on the initial energy E but is independent of the substance of the

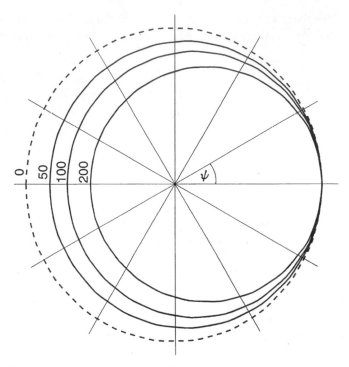

Figure 1-13. Compton scatter of a photon with an initial energy E. On collision with a loosely bound or free electron the photon is deflected by an angle ψ. It loses energy and keeps only the fraction E'/E. This fraction is represented by an ellipse in polar coordinates for a certain energy E (50, 100, and 200 keV). It may be compared with the outer dashed circle, representing conservation of the photon energy.

scatterer. It reaches a minimum for $\psi = 180°$ and decreases with the initial energy E.

In the wave picture, Compton scattering is controlled by a fixed wavelength shift $d\lambda = \lambda' - \lambda$. It amounts to

$$d\lambda = \lambda_c(1 - \cos\psi) \qquad (1\text{-}19)$$

where λ_c is a small constant called the Compton wavelength, which is defined by $\lambda_c = h/m_e c = 0.002426$ nm. The wavelength shift depends only on the deflection angle ψ and is independent of the wavelength λ itself.

The *intensity* of the scattered radiation shows a dependence on the initial energy E and the deflection angle ψ, as shown in Figure 1-14 [20]. Minimum intensity or scattering is achieved for a deflection around 90–100°. For that

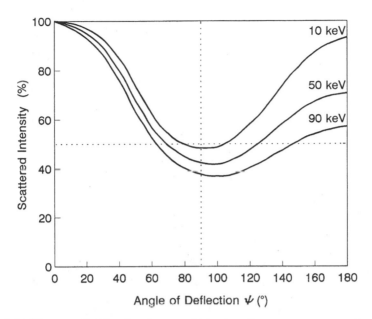

Figure 1-14. The normalized fraction of scattered *intensity* Compton scatter— as a function of the angle ψ by which the incident photon is deflected after collision with an electron. (From Woldseth [20].)

reason, a rectangular geometric arrangement of the X-ray tube, sample, and detector is generally chosen in X-ray spectrometry in order to minimize the inelastic scatter into the detector. Nevertheless, any primary radiation of an X-ray tube is scattered by the sample and is reproduced as a blank spectrum. In particular, the characteristic peaks of the tube anode give rise to the Rayleigh and Compton peaks in a sample spectrum. The corresponding *energies* can be calculated by equation (1-18) for $\psi = 90°$, independent of the sample itself. Their *intensities*, however, depend on the photon energy and moreover on the sample substance. Rayleigh scattering will increase if the energy of X-ray photons decreases or the mean atomic number of the scattering sample increases. Compton scattering, in contrast, will decrease if the photon energy decreases or the atomic number increases.

1.3.3. Total Attenuation

Photoelectric absorption and scattering jointly lead to the attenuation of X-rays in matter. The total mass-attenuation coefficient is composed additively by the photoelectric mass-absorption coefficient (τ/ρ) and the mass-scatter

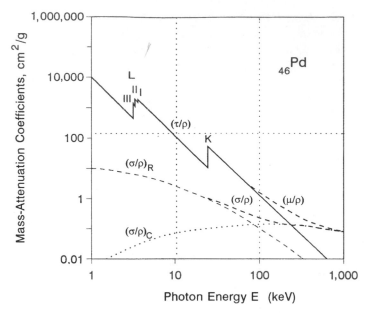

Figure 1-15. Photoelectric mass-absorption coefficient (τ/ρ) and mass-scatter coefficient (σ/ρ) as functions of the energy E of primary photons incident on palladium. The total mass-attenuation coefficient (μ/ρ) results from the sum $(\tau/\rho) + (\sigma/\rho)$, the latter term from the sum of Rayleigh and Compton scattering $(\sigma/\rho)_R + (\sigma/\rho)_C$. (From Veigele [21].)

coefficient (σ/ρ) according to

$$(\mu/\rho) = (\tau/\rho) + (\sigma/\rho) \tag{1-20}$$

Both fractions are shown in Figure 1-15 for the element palladium, as functions of the photon energy E [21]. The scatter coefficient (σ/ρ) is further divided into the Rayleigh and the Compton part. In contrast to the exponential decrease of (τ/ρ) with discontinuities at the absorption edges, the function (σ/ρ) varies more slightly and steadily. It decreases for Rayleigh scattering and increases for Compton scattering in the given energy range: for energies below 90 keV, the Rayleigh part is predominant; for energies above 90 keV, the Compton part. Both effects are relatively minor compared to the photoelectric absorption. But for energies above 200 keV, the compton effect becomes the decisive component of total attenuation. Similar conditions are valid for elements lighter or heavier than palladium. For light elements like carbon, the point of balance between Rayleigh and Compton scattering decreases to 10 keV; for heavy elements like lead, it increases to 150 keV [22].

It should be noted that the total mass-attenuation coefficient (μ/ρ) is mainly determined and equal to the photoelectric mass-absorption coefficient (τ/ρ) for lower photon energies (< 10 or even < 100 keV). In such cases, only one set of data is necessary for both quantities. For light elements like C and higher energies, however, both quantities (μ/ρ) and (τ/ρ) are different and so have to be distinguished.

In practice, energy-dependent attenuation is used to alter the spectrum of an X-ray beam. For that purpose, a thin metal sheet called a selective attenuation filter can be employed. It can easily be inserted into a beam path in order to reduce a particular spectral peak or an entire energy band with respect to other peaks or spectral regions.

1.4. DEFLECTION OF X-RAYS

In a homogeneous medium the X-ray beam just behaves like a light beam and follows a straight path on which the photons travel. But if the beam hits the boundary surface of a second medium, the incident X-ray beam will be deflected from its original direction. It can even be split, i.e., partly reflected into the first medium and partly refracted into the second medium.

1.4.1. Reflection and Refraction

In accord with Figure 1-16, the following rules are valid—

- The incident, the reflected, and the refracted beam span a plane that is normal to the boundary plane.
- The *glancing angles*[1] of the incident beam and the *reflected* beam are equal:

$$\alpha_1 = \alpha_1^* \qquad (1\text{-}21)$$

- The *glancing angles* of the incident beam and the *refracted* beam follow Snell's law:

$$v_2 \cos \alpha_1 = v_1 \cos \alpha_2 \qquad (1\text{-}22)$$

where v_1 and v_2 are the phase velocities of the beam in media 1 and 2, respectively. Phase velocity is the velocity at which the planes of constant

---

[1] Glancing angles are considered in X-ray optics. They are complements of the angles of incidence conventionally used in light optics.

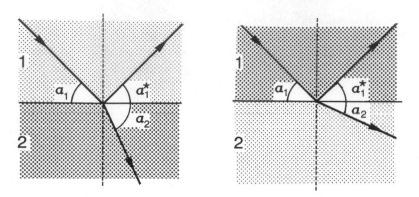

Figure 1-16. The incident, the reflected, and the refracted beam at the interface of two media 1 and 2. On the left, medium 2 is optically denser than medium 1 ($n_2 > n_1$); on the right, it is vice versa ($n_1 > n_2$).

phase, e.g., crests or troughs, propagate within a medium. It is dependent on the wavelength λ and the medium itself. *In vacuo*, the phase velocity has the value c (the light velocity) independent of the wavelength λ.

Division of equation (1-22) by the light velocity c results in

$$n_1 \cos \alpha_1 = n_2 \cos \alpha_2 \qquad (1\text{-}23)$$

where n_1 and n_2 are the absolute refractive indices of media 1 and 2, respectively, which are defined by

$$n_{1,2} = c/v_{1,2} \qquad (1\text{-}24)$$

Two cases can be distinguished, as demonstrated in Figure 1-16. If $n_2 > n_1$, i.e., if medium 2 is called optically denser than medium 1, the refracted beam in medium 2 will be deflected off the boundary. If $n_2 < n_1$, i.e. if medium 2 is optically thinner than medium 1, the refracted beam in medium 2 will be deflected toward the boundary.

The refractive index n is the decisive quantity and can be derived from the Lorentz theory assuming that the quasi-elastically bound electrons of the atoms are forced to oscillations by the primary radiation. As a result, the oscillating electrons radiate with a phase difference. By superposition of both radiations the primary one is altered in phase velocity. This alteration becomes apparent by a modified refractive index, deviating from the vacuum value $n_{\text{vac}} = 1$ by a small quantity δ.

If absorption cannot be neglected but must be taken into account, the refractive index n has to be written as a complex quantity. Conventionally, n is defined by

$$n = 1 - \delta - i\beta \qquad (1\text{-}25)$$

where i is the imaginary unit or the square root of -1. The real part of the refractive index accordingly is $n' = 1 - \delta$.

The *imaginary* component β is a measure of the attenuation already treated in Section 1.3. It can be expressed by

$$\beta = \frac{\lambda}{4\pi}\left(\frac{\mu}{\rho}\right)\rho \qquad (1\text{-}26)$$

The *real* component δ, called the decrement, measures the deviation of the real part n' of the refractive index from unity; δ determines the phase velocity v according to $v \approx c/(1 - \delta)$, which can even be greater than the light velocity.[2]

From theory it follows that δ can be written as [23]

$$\delta = \frac{N_A}{2\pi}r_e\rho\frac{1}{A}[f_0 + f(\lambda)]\lambda^2 \qquad (1\text{-}27)$$

where N_A = Avogadro's number = 6.022×10^{23} atoms/mol; r_e = the classical electron radius = 2.818×10^{-13} cm; ρ = the density (in g/cm³) of the respective element; A = atomic mass (in g/mol); λ = the wavelength of the primary beam; f_0 is a quantity that for X-rays is equal to the atomic number Z; and $f(\lambda)$ is a correction term [24] that is only decisive at and below the absorption edges ($E \leq E_j$ or $\lambda \geq \lambda_j$) and is generally negative. Consequently, δ includes some constants of matter and moreover strongly depends on the wavelength λ. This dependence is known as dispersion and is demonstrated in Figure 1-17 for the elements Cu and Au.

For primary X-rays—shorter in wavelength than the absorption edges— the f-values disappear and equation (1-27) can be simplified by

$$\delta = \frac{N_A}{2\pi}r_e\rho\frac{Z}{A}\lambda^2 \qquad (1\text{-}28)$$

---

[2] For X-rays with positive δ-values, the phase velocity v exceeds the light velocity c. This is possible since v is not a velocity at which a real signal can be transmitted. Only a signal velocity has to be smaller than the limiting light velocity.

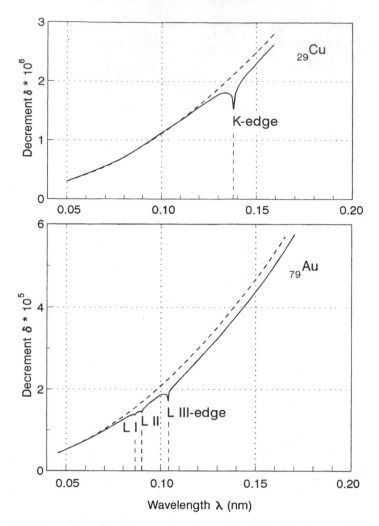

Figure 1-17. Dependence of the decrement δ on the wavelength λ for the elements Cu (above) and Au (below). The theory is based on forced oscillations of the atomic electrons—forced by the electromagnetic radiation of X-rays. At the "resonance" wavelengths or absorption edges, the decrement shows a strong variation. The asymptotic approximation for short wavelengths is represented by a dashed line. (Data from Stanglmeier [24].)

For X-rays, the quantity δ is mostly on the order of 10^{-6} and the real part $1 - \delta$ of the refractive index is somewhat smaller than 1. The minus sign in $1 - \delta$ reflects the fact that the bound electrons follow the excitant photons only slowly, i.e., with phase *opposition*. The small quantity of δ is due to the small amplitudes of the electrons' oscillations. Because of the high photon frequen-

Table 1-2. The Real Part δ and the Imaginary Part β of the Refractive Index n Calculated for Mo-$K\alpha$ X-rays with $\lambda = 0.071$ nm[a]

Medium	ρ (g/cm^3)	δ (10^{-6})	β (10^{-8})
Plexiglas	1.16	0.9	0.055
Glassy carbon	1.41	1.0	0.049
Boron nitride	2.29	1.5	0.090
Quartz glass	2.20	1.5	0.46
Aluminum	2.70	1.8	0.79
Silicon	2.33	1.6	0.84
Cobalt	8.92	5.6	19.8
Nickel	8.91	5.8	21.9
Copper	8.94	5.6	24.1
Germanium	5.32	3.2	18.7
Gallium arsenide	5.31	3.2	18.7
Tantalum	16.6	9.1	87.5
Platinum	21.45	11.7	138.2
Gold	19.3	10.5	129.5

[a] The various media of density ρ are listed in order of increasing (mean) atomic number.

cies corresponding to the short wavelengths of X-rays, only small amplitudes can occur.

In the X-ray region, the quantity β is even smaller than δ. Table 1-2 lists values of δ and β for some compounds and pure elements calculated for Mo-$K\alpha$ radiation. For compounds, solutions or mixtures δ and β have to be calculated according to the additive law already applied in equation (1-12):

$$\delta_{total} = \sum c_i \delta_i \qquad (1\text{-}29)$$

$$\beta_{total} = \sum c_i \beta_i \qquad (1\text{-}30)$$

Again, the c_i terms are the different mass fractions of the individual elements i with respective values δ_i and β_i.

1.4.2.   Total External Reflection

For X-rays, any medium is optically less dense than vacuum ($n' < n_{vac} = 1$) and any solid is optically less dense than air ($n' < n'_{air} \approx 1$). This normally results in a refracted beam that is deflected toward the boundary plane (Figure 1-16, right). If the respective glancing angle α_2 of the refractive beam becomes zero, the refracted beam will emerge tangentially to the boundary surface. Consequently, there is a minimum critical angle $\alpha_1 = \alpha_{\text{crit}}$ for which refraction is just possible. According to equation (1-23), this angle of incidence is determined by

$$\cos \alpha_{\text{crit}} = n_2 \qquad (1\text{-}31)$$

For angles α_1 even lower than α_{crit}, equation (1-23) gives no real value for the refraction angle α_2 since its cosine cannot be > 1. In this case no beam enters the second medium, but the boundary, like an ideal mirror, completely reflects the incident beam back into the first medium. This phenomenon is called *total reflection.*

The critical angle of total reflection can easily be calculated from equation (1-31). Since α_{crit} is small, its cosine can be approximated by

$$\cos \alpha_{\text{crit}} \approx 1 - \frac{\alpha_{\text{crit}}^2}{2} \qquad (1\text{-}32)$$

The combination with equation (1-25) leads to the simple relation

$$\alpha_{\text{crit}} \approx \sqrt{2\delta} \qquad (1\text{-}33)$$

Insertion of equation (1-28) gives the approximation

$$\alpha_{\text{crit}} \approx \frac{1.65}{E} \sqrt{\frac{Z}{A} \rho} \qquad (1\text{-}34)$$

where E has to be given in keV and ρ in g/cm^3 in order to get α_{crit} in degrees. As already mentioned for equation (1-28), this approximation is exactly valid for photon energies E above the decisive absorption edges of the material. Table 1-3 gives values for different media and photon energies, which all lie between $0.04°$ and $0.6°$.

In the range of total reflection, extensive calculations have to be carried out on complex refraction angles α_2 with a real and an imaginary part. Nevertheless, for our purposes it is possible to make this task a lot easier, as simple approximations can be applied for the small glancing angles considered in X-ray optics.

In accord with Figure 1-16 (right), X-rays are assumed to run through a vacuum and then to strike a medium at an angle α_1. In this case, the angle α_2 of the refracted beam has to be considered complex. From Snell's law (1-23) we have

$$\alpha_2 \approx \sqrt{\alpha_1^2 - 2\delta - 2i\beta} \qquad (1\text{-}35)$$

Table 1-3. Critical Angle α_{crit} of Total Reflection, Calculated for Various Media and X-rays of Different Photon Energies

	α_{crit} at Photon Energy of:		
Medium	8.4 keV (degree)	17.44 keV (degree)	35 keV (degree)
Plexiglas	0.157	0.076	0.038
Glassy carbon	0.165	0.080	0.040
Boron nitride	0.21	0.10	0.050
Quartz glass	0.21	0.10	0.050
Aluminum	0.22	0.11	0.054
Silicon	0.21	0.10	0.051
Cobalt	0.40	0.19	0.095
Nickel	0.41	0.20	0.097
Copper	0.40	0.19	0.095
Germanium	0.30	0.15	0.072
Gallium arsenide	0.30	0.15	0.072
Tantalum	0.51	0.25	0.122
Platinum	0.58	0.28	0.138
Gold	0.55	0.26	0.131

where δ and β belong to the complex refractive index n of the medium. The real and imaginary components of this angle, α_2' and α_2'', respectively, can be written as (see, e.g., Stanglmeier [24] and Blochin [25])

$$\alpha_2'^2 = \tfrac{1}{2}\left[\sqrt{|(\alpha_1^2 - 2\delta)^2 + (2\beta)^2|} + (\alpha_1^2 - 2\delta)\right] \tag{1-36}$$

$$\alpha_2''^2 = \tfrac{1}{2}\left[\sqrt{|(\alpha_1^2 - 2\delta)^2 + (2\beta)^2|} - (\alpha_1^2 - 2\delta)\right] \tag{1-37}$$

Both components are represented in Figure 1-18 for Mo-$K\alpha$ X-rays striking a flat silicon substrate. The real component α_2' is dominant in the range above the critical angle α_{crit} and is asymptotically equal to α_1 for large angles. The imaginary component α_2'' is decisive for angles below α_{crit} and is asymptotically equal to α_{crit} for small angles. Both components become equal at the critical angle and amount to $\sqrt{\beta}$, which is extremely small. Moreover, the product of both components always equals β independent of the given glancing angle of incidence α_1:

$$\alpha_2' \cdot \alpha_2'' = \beta \tag{1-38}$$

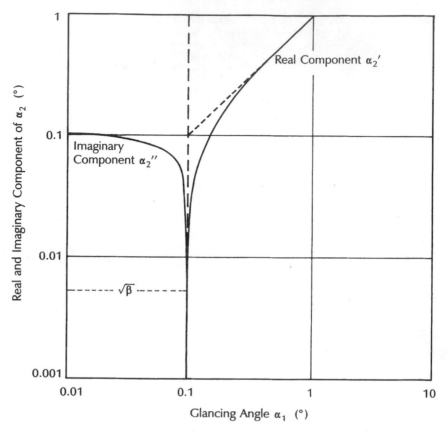

Figure 1-18. Real and imaginary component in a double-logarithmic plot, determining the angle α_2 of the refracted X-ray beam in dependence on the glancing angle α_1 of the incident beam. Calculation for Mo-$K\alpha$ X-rays striking a flat silicon substrate above and below the critical angle of total reflection that amounts to $0.1°$.

This relationship can easily be verified by multiplication of equations (1-36) and (1-37). It is important for subsequent calculations in Section 2.4.1.

Two important quantities characterize total reflection:

- The reflectivity R, which is increased to 100% below the critical angle α_{crit}
- The penetration depth z_n, which is reduced to a few nanometers in this case

Both quantities can be calculated from the theory of a harmonic and plane electromagnetic wave, as discussed in the following subsections.

1.4.2.1. Reflectivity

The reflectivity is defined by the intensity ratio of the reflected beam and the incident beam and can be derived from the Fresnel formulas. These well-known formulas connect the vectors of the electromagnetic field of the reflected and the transmitted beam with those of the incident beam [26]. For the grazing incidence considered here, the amplitudes E_1, E_1^*, and E_2 of the electric field vectors are expressed by the simple formulas

$$\frac{E_1^*}{E_1} = \frac{\alpha_1 - \alpha_2}{\alpha_1 + \alpha_2}$$

$$\frac{E_2}{E_1} = \frac{2\alpha_1}{\alpha_1 + \alpha_2}$$

(1-39)

These formulas are valid independent of the polarization of the incident beam because of the assumed small angles α_1 and α_2. The reflectivity and transmissivity follow from these formulas after the absolute magnitude is squared.

The reflectivity R is given by

$$R = \left| \frac{\alpha_1 - \alpha_2}{\alpha_1 + \alpha_2} \right|^2$$

(1-40)

With the help of the components α_2' and α_2'' of the complex angle α_2, the reflectivity can be calculated:

$$R = \frac{(\alpha_1 - \alpha_2')^2 + \alpha_2''^2}{(\alpha_1 + \alpha_2')^2 + \alpha_2''^2}$$

(1-41)

Three highly useful approximations result

$$\alpha_1 \ll \alpha_{crit}: \qquad R \simeq 1 - \sqrt{\frac{2\beta}{\delta\delta}}\alpha_1$$

$$\alpha_1 = \alpha_{crit}: \qquad R \approx \frac{\delta + \beta - \sqrt{(2\beta\delta)}}{\delta + \beta + \sqrt{(2\beta\delta)}}$$

(1-42)

$$\alpha_1 \gg \alpha_{crit}: \qquad R \simeq \frac{\delta^2}{4\alpha_1^4}$$

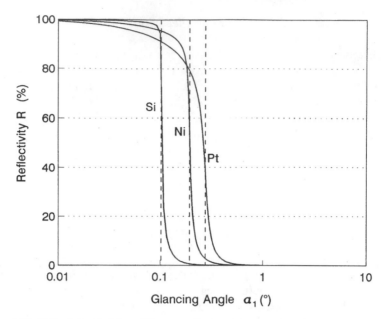

Figure 1-19. Reflectivity R of three different media dependent on the glancing angle α_1 of X-rays. The curves were calculated for X-rays of Mo-$K\alpha$. Reflection below a critical angle α_{crit} is called *total reflection*. These angles are determined by the point of inflection of the curves and are marked by dashed lines.

The dependence of the reflectivity R on the glancing angle α_1 is demonstrated in Figure 1-19. The effect of total reflection is shown for three different elements. For glancing angles of $1°$ or more, the reflectivity is below 0.1%, independent of absorption, and can be neglected generally. Around the critical angle α_{crit}, the reflectivity rises to high values. However, the rise to 100% is not steplike but more or less gradual, dependent on the absorption or attenuation quantity β. The angle α_{crit} determines the point of inflection of the curves. For a less absorbing medium like silicon, the reflectivity shows the most distinct transition. For this reason silicon or quartz glass and even Plexiglas are used as sample carriers for TXRF.

The curves of Figure 1-19 were calculated for X-rays of Mo-$K\alpha$ with a photon energy of 17.44 keV. For higher energies, the α_{crit}-values are decreased according to the $1.65/E$ term in equation (1-34) and consequently the curves are shifted to the left. For lower energies, they are shifted to the right (see Figure 3-4). Table 1-4 lists the corresponding reflectivity values R_{crit} calculated by equation (1-42).

Table 1-4. Reflectivity R_{crit} of Various Media at the Critical Angle of Total Reflection, Calculated for X-rays of Different Photon Energies

Medium	R_{crit} at Photon Energy of:		
	8.4 keV (%)	17.44 keV (%)	35 keV (%)
Plexiglas	87.9	93.2	94.8
Glassy carbon	88.4	93.9	95.0
Boron nitride	87.6	93.3	94.6
Quartz glass	73.4	85.5	91.4
Aluminum	69.7	82.9	90.3
Silicon	67.3	81.5	89.5
Cobalt	37.4	59.1	75.2
Nickel	37.0	58.1	74.9
Copper	66.9	56.1	82.7
Germanium	62.3	51.2	69.7
Gallium arsenide	62.4	51.1	69.5
Tantalum	49.3	42.9	63.4
Platinum	45.3	39.4	60.2
Gold	44.8	38.7	59.5

1.4.2.2.   Penetration Depth

The penetration depth is defined by that depth of a homogeneous medium a beam can penetrate while its *intensity* is reduced to $1/e$, or 37% of its initial value. This depth z_n, which is normal to the boundary of the medium, follows the equation

$$z_n \approx \frac{\lambda}{4\pi}\frac{1}{\alpha_2''} \tag{1-43}$$

Again three approximate values can be given:

$$\alpha_1 \ll \alpha_{crit}: \quad z_n \approx \frac{\lambda}{4\pi}\frac{1}{\sqrt{2\delta}}$$

$$\alpha_1 = \alpha_{crit}: \quad z_n \approx \frac{\lambda}{4\pi}\frac{1}{\sqrt{\beta}} \tag{1-44}$$

$$\alpha_1 \gg \alpha_{crit}: \quad z_n \simeq \frac{\lambda}{4\pi}\frac{\alpha_1}{\beta}$$

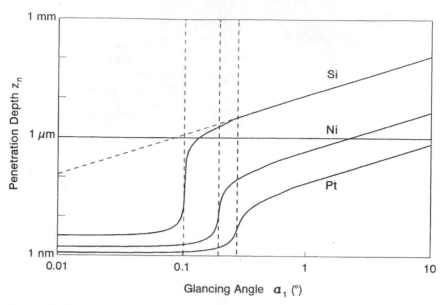

Figure 1-20. Penetration depth z_n of X-rays hitting three different media at a variable glancing angle α_1. The curves were calculated for X-rays of Mo-$K\alpha$ with a photon energy of 17.44 keV. The critical angles are marked by dashed vertical lines. The dashed oblique straight line represents the penetration depth in a roughened Si surface for which total reflection disappears.

Figure 1-20 shows a double-logarithmic presentation of the penetration depth dependent on the glancing angle α_1 for the three elements already considered in Figure 1-19. For angles above 0.5°, the penetration depth linearly decreases with the glancing angle and the depth values are on the order of 0.1–10 μm. At the critical angle α_{crit}, the penetration depth drastically decreases especially for nonabsorbing media like silicon. Below this critical angle, the penetration depth reaches a constant level of only a few nanometers.

Of course, the effect of total reflection only appears when the medium is flat and smooth. For a rough surface, total reflection disappears. The penetration depth linearly decreases with the glancing angle even below the critical angle, as is demonstrated in Figure 1-20 for silicon.

The curves of Figure 1-20 were calculated for the photon energy of the chosen Mo-$K\alpha$ radiation. Figure 1-21 shows the influence of different photon energies on the penetration depth—here of silicon. The points of inflection shift to lower critical angles with increasing photon energy. Furthermore, the curves are stretched to higher depth values for normal reflection while the

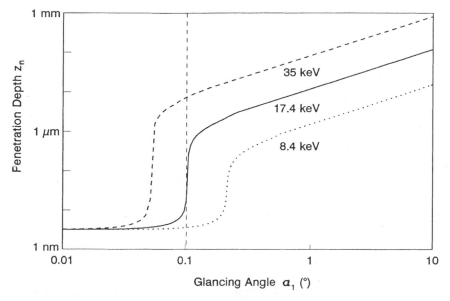

Figure 1-21. Penetration depth z_n of X-rays striking on silicon at a variable glancing angle α_1. The curves were calculated for three different photon energies. The dashed vertical line signifies the respective critical angle.

Table 1-5. Minimum Penetration Depth z_0 and Penetration z_{crit} of Various media, Calculated for X-Rays of Different Photon Energies

Medium	Minimum z_0 (nm)	z_{crit} at Photon Energy of:		
		8.4 keV (nm)	17.44 keV (nm)	35 keV (nm)
Plexiglas	4.3	132	241	319
Glassy carbon	4.1	130	255	311
Boron nitride	3.2	97	188	238
Quartz glass	3.2	42	83	146
Aluminum	3.0	33	64	116
Silicon	3.2	32	62	115
Cobalt	1.7	6.6	12.7	24
Nickel	1.7	6.4	12.1	23
Copper	1.7	16.8	11.5	22
Germanium	2.2	18.8	13.1	25
Gallium arsenide	2.2	18.8	13.0	24
Tantalum	1.3	7.3	6.0	11.4
Platinum	1.2	5.8	4.8	9.1
Gold	1.2	6.0	5.0	9.4

depth values for total reflection remain constant at

$$z_0 \simeq 3.424 \sqrt{\frac{A}{Z}\frac{1}{\rho}} \tag{1-45}$$

where ρ has to be given in g/cm^3 to get z_0 in nm. This minimum is a material constant that is independent of the photon energy E and is listed in Table 1-5 for various media. Table 1-5 also gives penetration depths z_{crit} calculated for the critical angle α_{crit} and three different photon energies according to equation (1-44).

REFERENCES

1. Moseley, H.G.J. (1913). *Philos. Mag.* [6] **26**, 1024; (1914). **27**, 703.

2. Compton, A.H. (1923). *Philos. Mag.* [6] **45**, 1121.

3. Yoneda, Y., and Horiuchi, T. (1971). *Rev. Sci. Instrum.* **42**, 1069.

4. Aiginger, H., and Wobrauschek, P. (1974). *Nucl. Instrum. Methods* **114**, 157.

5. Aiginger, H., and Wobrauschek, P. (1975). *Anal. Chem.* **47**, 852.

6. Knoth, J., and Schwenke, H. (1978). *Fresenius' Z. Anal. Chem.* **291**, 200.

7. Schwenke, H., and Knoth, J. (1982). *Nucl. Instrum. Methods* **193**, 239.

8. Becker, R.S., Golovchenko, J.A., and Patel, J.R. (1983). *Phys. Rev. Lett.* **50**, 153.

9. Kiessig, H. (1931). *Ann. Phys.* (Leipzig) **10**, 769.

10. DuMond, J., and Youtz, J.P. (1940). *J. Appl. Phys.* **11**, 357.

11. Boumans, P.W.J.M., and Klockenkämper, R., eds. (1989). *Spectrochim. Acta* **44B**, 433–549.

12. Boumans, P.W.J.M., Wobrauschek, P., and Aiginger, H., eds. (1991). *Spectrochim Acta* **46B**, 1313–1436.

13. Boumans, P.W.J.M., and Prange, A., eds. (1993). *Spectrochim. Acta* **48B**, 107–299.

14. Taniguchi, K., ed. (1995). *Adv. X-Ray Chem. Anal. Jpn.* **26s**, 1–206.

15. Bertin, E.P. (1975). *Principles and Practice of Quantitative X-Ray Fluorescence Analysis*, 2nd ed. Plenum Press, New York.

16. Jenkins, R. (1988). *X-Ray Fluorescence Spectrometry*, Chemical Analysis Series. Wiley (Interscience), London.

17. Tertian, R., and Claisse, F. (1982). *Principles of Quantitative X-Ray Fluorescence Analysis*. Heyden, London.

18. Williams, K.L. (1987). *An Introduction to X-Ray Spectrometry*. Allen & Unwin, London.

19. Krieger, H., and Petzold, W. (1989). *Strahlenphysik, Dosimetrie und Strahlenschutz*, Vol. 2. Teubner, Stuttgart.

20. Woldseth, R. (1973). *X-Ray Energy Spectrometry*, Kevex Corp., Burlingame, California.

21. Veigele, W.J. (1973). *At. Data Tables* **5**.

22. Author collective (1986). *X-Ray Data Booklet* (D. Vaughan ed.). Lawrence Berkeley Laboratory, Berkeley, California.

23. James, R.W. (1967). *The Optical Principles of the Diffraction of X-Rays*. Cornell University Press, Ithaca, N.Y.

24. Stanglmeier, F. (1990). *Forschungszentrum Jülich*. Rep. No. 2346 (Dissertation, Technische Hochschule, Aachen).

25. Blochin, M.A. (1957). *Physik der Röntgenstrahlen*, VEB Verlag Technik, Berlin.

26. Born, M., and Wolf, E. (1980). *Principles of Optics*. Pergamon Press, London; 6th ed., reprinted 1993.

PRINCIPLES OF TOTAL-REFLECTION XRF

Our consideration of the fundamentals of X-ray fluorescence so far has treated the nature of X-rays and their production, attenuation, and deflection. These fundamentals are sufficient as a basis of conventional XRF spectrometry. However, some additional premises are required to elucidate the peculiarities of total-reflection X-ray fluorescence (TXRF) and related methods. These premises especially concern the phenomena of interference and standing waves. While conventional excitation takes place in a uniform field of X-radiation, TXRF occurs in the inhomogeneous field of standing waves in front of optically flat substrates or within thin layers on such substrates. The occurrence of such a field consequently has to be considered in some detail.

2.1. INTERFERENCE OF X-RAYS

The phenomena of interference result from the superposition of two beams (double-beam interference) or even more than two beams (multiple-beam interference). They are generally explained in the wave picture. In the region of superposition the resulting wave field can show a pattern with maxima and minima. These fluctuations will be highly distinct if two superimposing waves are monochromatic and coherent, i.e., if they have the same wavelength and a fixed phase difference. If the phase difference is an odd multiple of π, the amplitudes are subtracted to a minimum. This kind of interference is called destructive, and points with minima are called nodes. On the other hand, interference is called constructive if the phase difference is an even multiple of π and the amplitudes sum up to a maximum. The corresponding points are called antinodes. Nodes and antinodes can be extended to nodal and anti-nodal lines or planes, respectively.

A most simple way to produce interference is the superposition of two beams propagating on the same straight line. This two-beam interference can be effected by reflection of X-rays at the upper and lower boundaries of a thin layer deposited on a thick substrate. It can become multiple-beam interference if a multilayer is chosen with a reflection at various boundaries. Furthermore, the scattering of X-rays at different atoms of a crystal can lead to multiple-beam interference. The specific behavior can be explained by reflection at the various lattice planes, which results in the well-known Bragg law. This interference of coherently scattered X-rays is often called *diffraction*.

2.1.1. Double-Beam Interference

A thin homogeneous layer may be deposited on a thick and flat substrate. If a monochromatic X-ray beam hits this layer at grazing incidence, double-beam interference can be observed. Such experiments require a glancing angle of about 0.1° and a layer thickness of 10 nm to 1 μm. Furthermore, a beam width of some 10 μm and a layer size of some 10 mm are required for a wide superposition.

Figure 2-1 depicts the paths of X-rays within the three media denoted by subscripts 0, 1, and s: the first medium, from which the X-rays are coming, is assumed to be vacuum or air; the second medium is a thin plane-parallel layer of thickness d; and the third medium is a thick substrate. Two cases, A and B, can be distinguished. In case A, the layer is optically denser than the substrate ($n'_1 > n'_s$, where n' is the real part of the refractive index n; see Section 1.4.1). In case B, the layer is optically thinner than the substrate ($n'_1 < n'_s$).

In case A, the glancing angle α_1 is smaller than α_0 and α_s is even smaller than α_1, so that total reflection is possible at *both* interfaces: air–layer and layer–substrate. The first happens at and below a small angle determined by $\alpha_{01} = \sqrt{2\delta_1}$ when α_1 becomes zero or imaginary (see Section 1.4.2). The last happens below the greater angle $\alpha_{0s} = \sqrt{2\delta_s}$ when α_1 becomes $\sqrt{2(\delta_s - \delta_1)}$ and α_s becomes zero or imaginary. In case B, α_1 is likewise smaller than α_0 but α_s is generally greater than α_1, so that total reflection is only possible at the upper, air–layer interface but not at the lower, layer–substrate interface. The first again happens below the angle $\alpha_{01} = \sqrt{2\delta_1}$ when α_1 becomes zero or imaginary.

Of course, the actual glancing angles for grazing incidence of X-rays are much smaller than those illustrated in Figure 2-1. Moreover, multiple zigzag reflections not considered here are generally possible. Finally, the two boundaries assumed to be perfectly smooth usually have a certain roughness. Nevertheless, the idealized model leads to a first approximative description of the interference phenomena.

For that purpose, the phase difference of the two reflected beams coming from the lower and upper boundary has to be determined. These two beams have a path difference Δ given by $|AB| + |BC| - |CD|$:

$$\Delta = 2d \sin \alpha_1 \qquad (2\text{-}1)$$

The path difference can be approximated after use of equation (1-35):

$$\Delta \approx 2d \sqrt{\alpha_0^2 - 2\delta_1} \qquad (2\text{-}2)$$

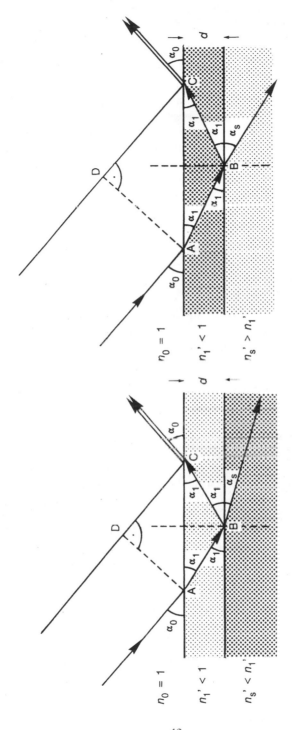

Figure 2-1. Paths of X-rays striking a thin layer of thickness d deposited on a thick substrate. Case A, left: the layer is optically denser than the substrate ($n_0 = 1 > n_1' > n_s'$). Case B, right: the layer is optically thinner than the substrate ($n_0 = 1 > n_1' < n_s'$).

43

where α_0 is the glancing angle of the incoming beam, and δ_1 is the decrement of the layer. The phase difference φ of the two reflected beams is given by $(2\pi/\lambda)\cdot\Delta$, at least in case A. In case B, the amount of π still has to be added due to a phase jump at the optically denser substrate.

As already mentioned, constructive or destructive interference occurs if the phase difference φ is an even or odd multiple of π, respectively. If k represents this even or odd integer, the condition can be written as

$$\frac{4\pi}{\lambda}d\sqrt{\alpha_0^2 - 2\delta_1} \approx k\pi \qquad (2\text{-}3)$$

This relation is fulfilled for two sets of glancing angles $\alpha_0 = \alpha_k$. The maxima occur for

$$\alpha_k^2 \approx \alpha_{01}^2 + \left(k\frac{\lambda}{2d}\right)^2 \qquad (2\text{-}4a)$$

and the minima for

$$\alpha_k^2 \approx \alpha_{01}^2 + \left(\frac{2k+1}{2}\frac{\lambda}{2d}\right)^2 \qquad (2\text{-}4b)$$

where α_{01} is the critical angle of total reflection for the layer material.

The angles α_k determine the directions in which the two reflected beams are reinforced or annihilated. They lead to maxima and minima for the reflectivity of the layered substrate. Cases A and B differ by exchange of maxima and minima at the same k-value.

The reflectivity can be determined by measuring the intensity of the reflected beams at an angle of $\alpha_0^* = \alpha_0$. The ratio of this value to the intensity of the incoming beam gives the reflectivity R. This quantity can also be calculated theoretically. It depends on the two values of reflectivity—that of the layer, R_{01}, and that of the substrate, R_{1s}. These two values result from equation (1-41) and finally lead to the reflectivity R of the layered substrate [1,2]:

$$R = \frac{R_{01} + 2\sqrt{(R_{01}R_{1s})}\exp(-\varphi'')\cos\varphi' + R_{1s}\exp(-2\varphi'')}{1 + 2\sqrt{(R_{01}R_{1s})}\exp(-\varphi'')\cos\varphi' + R_{01}R_{1s}\exp(-2\varphi'')} \qquad (2\text{-}5)$$

where φ' and φ'' are the real and imaginary components, respectively, of the complex phase difference φ between the two reflected beams; φ is determined by

$$\varphi = \frac{4\pi}{\lambda}d\alpha_1 \qquad (2\text{-}6)$$

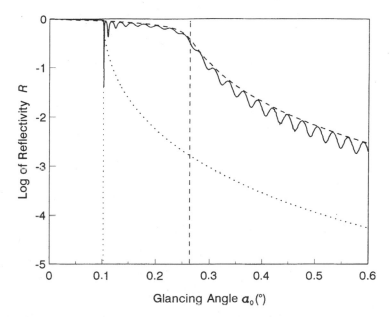

Figure 2-2. Reflectivity R of a thin layer deposited on a thick substrate in dependence on the glancing angle α_0. Case A: a 70 nm Si-layer on a Au-substrate. Curves were calculated for Mo-$K\alpha$ radiation with $\lambda = 0.071$ nm. The dotted curve indicates the reflectivity of the layer material Si; the dashed curve, that of the substrate material Au. The dotted and dashed vertical lines represent the critical angles for the layer Si and the substrate Au, respectively. (After de Boer [3].)

where α_1 is the complex glancing angle of the refracted beam given by equation (1-35). As already mentioned, the value π has to be added if case B takes place instead of case A.

The reflectivity R ultimately is a function of the glancing angle α_0 of the incoming beam. This function is demonstrated in Figure 2-2 (after de Boer [3]) for an example of case A and in Figure 2-3 (after de Boer and van den Hoogenhof [4]) for a chosen case B. The semilogarithmic plot spans 5 orders of magnitude for the reflectivity. The value of R is roughly approximated by the greater value of R_{01} and R_{1s}. Maxima and minima appear just above the critical angle α_{01} of the respective layer at the angles α_k, in accord with equation (2-4). They are called Kiessig maxima and minima because H. Kiessig was the first to observe equivalent fringes for the reflectivity of a layer [5]. The first minimum is extremely marked in case A, which is discussed in Section 2.2.2.

The distances between the maxima or minima asymptotically follow with an angular period:

$$\Delta\alpha = \lambda/2d \qquad (2-7)$$

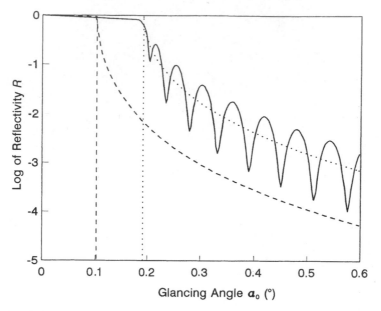

Figure 2-3. Reflectivity R of a thin layer deposited on a thick substrate as a function of the glancing angle α_0. Case B: a 30 nm Co-layer on a Si-substrate. Curves were calculated for Mo-$K\alpha$ radiation with $\lambda = 0.071$ nm. The dotted curve indicates the reflectivity of the layer material Co; the dashed curve, that of the substrate material Si. The dotted and dashed vertical lines represent the critical angles for the layer Co and the substrate Si, respectively. (After de Boer and van den Hoogenhof [4]).

Consequently, the extrema of a thinner layer show a greater period, e.g., for case B compared to case A. Reflectivity measurements can be carried out in order to determine the thickness of a layer. The thickness can be estimated by comparison of the measured and calculated curves of the reflectivity. This method, however, cannot determine the chemical composition of an unknown layer or substrate.

2.1.2. Multiple-Beam Interference

In addition to the great interest in single thin layers, there is much interest in double layers (or bilayers) and even in more complicated multilayers.

An ideal multilayer consists of different layers, each being isotropic and homogeneous. Their boundaries are plane parallel to each other and to the surface of the substrate on which the layers are deposited. The properties of each layer are considered constant throughout each plane, parallel to the surface. Such a system is generally called a *stratified medium* or a medium with

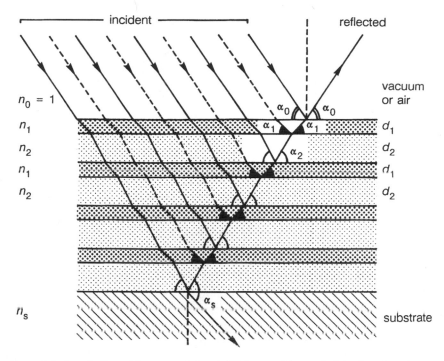

Figure 2-4. A periodic multilayer consisting of N bilayers with a period $d = d_1 + d_2$ and a total thickness $D = Nd$, deposited on a thick flat substrate. The layers with index 1 are the reflectors; the layers with index 2 are the spacers. After coherent scattering, several reflected X-ray beams can interfere with each other.

a *stratified structure*. The most important structures are periodic, composed of a stack of N bilayers. Figure 2-4 shows a succession of bilayers consisting of alternately lower and higher refractive indices n_1 and n_2, respectively, and with thicknesses d_1 and d_2. The sum $d = d_1 + d_2$ is the thickness of each bilayer. With respect to their specific function, the first layer is called the *reflector*; the second, the *spacer*. Details on available multilayers are given in Section 3.3.2.

A special kind of periodic multilayers called Langmuir–Blodgett layers consists of multiple smectic soap films with a pseudocrystalline structure. Even flat plates of true anorganic or organic crystals can be regarded as periodic multilayers. In this case, the atoms of a crystal-lattice plane build the one "layer," and the intermediate space between two successive crystal planes build the other "layer," all the layers being parallel to the surface plane. The interplanar spacing d between two neighboring crystal planes corresponds to the aforementioned sum $d_1 + d_2$.

A special feature of all these periodic multilayers is the *Bragg reflection*. Figure 2-4 shows X-ray beams incident under a glancing angle α_0 and following polygonal courses within a multilayer. At any interface, the beams are refracted and reflected under an angle α_1 or α_2, respectively. In total, reflection results from X-rays being scattered coherently from crystal planes or layers and interfering with each other. Several beams can leave the multilayer together under the glancing angle α_0 of reflection. X-rays of certain wavelengths λ are reinforced in intensity by constructive interference, while other X-rays are more or less annihilated by destructive interference. Constructive interference will happen if the path difference of X-rays reflected from two neighboring planes or reflectors is an integer of λ. Neglecting absorption and refraction, Bragg's law can be written as a simple equation

$$2d \sin \alpha_m \approx m \lambda \qquad (2\text{-}8)$$

where m is an arbitrary integer; λ, the wavelength of X-rays; and α_m, the glancing angles of incidence and reflection of these X-rays. The law implies that a strong reflection for a given wavelength is possible for certain angles α_m or in certain *directions* α_m, as is demonstrated in Figure 2-5. The different possible reflections are called reflections of the mth order.

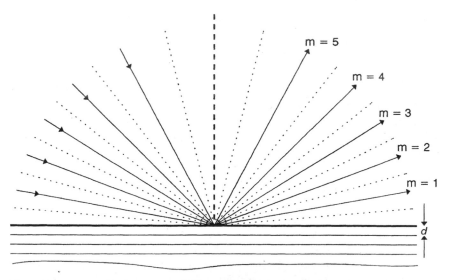

Figure 2-5. Bragg reflection at a periodic multilayer or crystal with a period d. Maxima of reflection appear for certain directions (——); minima are in between them (......). This illustration is valid for X-rays of Mo-$K\alpha$ reflected at a LiF crystal with $d = 0.20$nm.

A more stringent equation for multilayers [6] accounts for the refraction and absorption of the incoming beam within the reflector, 1, and the spacer, 2:

$$2d_1 \sin \alpha_1 + 2d_2 \sin \alpha_2 = m\lambda \qquad (2\text{-}9)$$

where d_1 and d_2 are the thicknesses of the reflector and spacer, respectively; α_1 and α_2 are the complex refraction angles determined by equation (1-35). For X-rays, the original Bragg equation (2-8) is a good approximation of the above stringent relation.

Natural crystals and pseudocrystals have long been used as Bragg monochromators. They constitute the cornerstone of wavelength-dispersive spectrometry. In recent years, periodic multilayers have been used as well. They are especially suitable for the long-wavelength region and consequently for the determination of light elements with low atomic numbers Z. In TXRF instruments, multilayers are employed as monochromators for short wavelengths (see Section 3.3.2). Furthermore, such multilayers can serve as test samples to be examined by TXRF or related methods. In this case, the decisive glancing angles are small and connected with the range of total reflection. Bragg's law in equation (2-8) can be rearranged after a simple approximation:

$$\alpha_m^2 \approx \alpha_{01}^2 + \left(m \frac{\lambda}{2d} \right)^2 \qquad (2\text{-}10)$$

where α_m values are the Bragg angles for maximum reflectivity of the mth order; α_{01} is a critical angle for total reflection of the multilayer.

As was already mentioned for a single layer, the reflectivity can be measured or even calculated. A suitable algorithm for a multilayer is given later (in Section 2.4.3). The results are somewhat complicated, but their depiction, e.g., in Figure 2-6 is not. The reflectivity curve is shown as a semilogarithmic plot (after Huang and Parrish [7]) dependent on the glancing angle α_0. The range of total reflection with $R \approx 1$ and the first and second Bragg maxima can clearly be assigned to the chosen periodic multilayer. Besides, there are several smaller oscillations that may be characterized as Kiessig maxima and minima.

The multilayer can obviously be considered as a single layer of thickness $D = Nd$. The Kiessig maxima occur for angles α_k in accord with equation (2-4a), where d is to be replaced by D. The comparison with equation (2-10) reveals the first Bragg reflection as the Nth Kiessig maximum. In general, Bragg reflections are identical with certain Kiessig maxima. They appear if k equals the number N of bilayers or an m-fold of this value ($k = Nm$). The Bragg maxima significantly exceed the "normal" Kiessig maxima by about 1 order of magnitude. A higher degree of resonance or symmetry is obviously reached in these cases (considered further in Section 2.2.3).

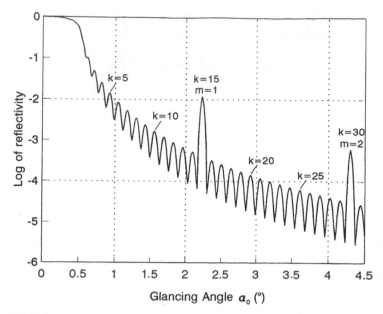

Figure 2-6. Reflectivity of a multilayer dependent on the glancing angle α_0. The multilayer was assumed to consist of 15 bilayers of Pt and Co, each of them 1.9 nm or 0.2 nm thick, respectively. Reflection is observed for a Cu-$K\alpha$ beam with photons of 8.047 keV. The lower Kiessig maxima can easily be distinguished from the higher Bragg maxima with $m = 1$ and $m = 2$. (After Huang and Parrish [7].)

The Bragg maxima usually decrease as the order of reflection becomes higher. There is, however, one important exception. If the thickness ratio d_2/d_1 is an integer i, the order $(i + 1)$ of reflection will vanish [6].

2.2. STANDING WAVE FIELDS

The interference as treated in Section 2.1 results from the superposition of two or more beams propagating in the same direction. Interference, however, is even possible for beams propagating in different directions. If the region of superposition is extensive, a total wave field can be observed. The resulting wave pattern can propagate with a certain velocity in a certain direction. But it can also be stationary in a direction where it does not move at all. This phenomenon is called a *standing wave*. It can be described as simple oscillations with locally dependent amplitudes. The locally fixed minima are called nodes and the maxima are called antinodes, as mentioned previously.

A very simple way to produce a standing wave is the superposition of a wide incoming and a wide reflected wave. This can be accomplished in front of a totally reflecting surface of a thick substrate, within a thin layer on such a substrate, or even within a multilayer. In any case, standing waves are formed that have a fundamental importance for TXRF, where excitation is performed either in front of totally reflecting flat substrates or within highly reflecting layers.

Real substrates or layers unfortunately are not ideally flat and smooth but show at least a certain microroughness. This should and can be taken into account and included in the calculations. For simplicity and clarity, however, this phenomenon will be ignored in this section. Substrates or layers are assumed to be optically flat in the nanometer range. For the practical examples of Chapter 4, however, even a rough surface will be taken into consideration.

2.2.1. Standing Wave Fields in Front of a Thick Substrate

When a plane monochromatic wave hits a flat reflecting surface of a thick substrate, interference occurs in the triangular section above the surface where the two waves cross each other. Figure 2-7 shows an instantaneous picture of the interference pattern, with a gray scale for the strength of the electromagnetic field. This pattern moves from the left to the right parallel to the surface. In all directions normal to the surface, however, standing waves can be observed. Parallel to the surface, there are nodal lines with zero amplitude

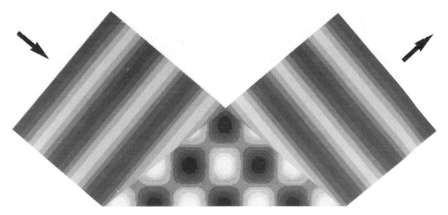

Figure 2-7. Interference of the incoming and the reflected X-ray waves in the triangular region above a flat and thick reflecting substrate. The strength of the electromagnetic field is represented on a gray scale with instantaneous crests (white) and troughs (black). In the course of time, the pattern moves from the left to the right.

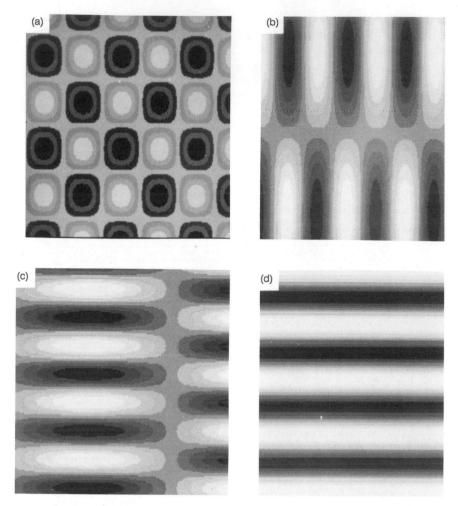

Figure 2-8. Interference patterns for different glancing angles α of the incident plane wave: (a) $\alpha = 45°$; (b) $\alpha = 10°$; (c) $\alpha = 80°$; (d) $\alpha = 90°$.

(gray) that follow one another at a constant distance. The antinodal lines with extreme amplitude are exactly between the nodal lines (crests are white and troughs are black).

The pattern of course depends on the angle by which the primary beam is incident. Four examples are given in Figure 2-8. (a) For $\alpha = 45°$, the nodal lines are the same distance apart as the crests and troughs are, so that a checkered pattern arises. (b) For glancing angles $\alpha < 45°$, the nodal lines are pulled apart while the crests and troughs of the antinodal lines move closer together. This pattern is formed in X-ray optics because of the small angles at total reflection. (c) For angles $\alpha > 45°$, the nodal lines are compressed while the crests and troughs move further apart. (d) At normal incidence, i.e., $\alpha = 90°$, the crests and troughs are horizontally fixed and nodal lines follow one another at a distance of half a wavelength. This pattern represents the most simple and familiar picture of a standing wave as is known from light optics.

For a mathematical description of the pattern some assumptions have to be made. The substrate shall first be regarded as a totally reflecting mirror with a reflectivity $R = 100\%$. The xy-plane shall be the surface plane; the z-axis, a normal of that plane; and α, the glancing angle of an X-ray beam incident from the vacuum and reflected at the substrate. The beam shall represent a monochromatic plane wave with wavelength λ, velocity c, and amplitude A_0 of the electric-field strength. In front of the substrate, the incident and reflected beams cross at an angle 2α. Based on these conditions, the interference can be described by the electric-field strength A_{int} according to

$$A_{int} = 2A_0 \sin(k_0 z \sin \alpha) \cos(k_0 ct - k_0 x \cos \alpha + \varphi) \qquad (2\text{-}11)$$

where $k_0 - 2\pi/\lambda$ is the wavenumber; t, the time; and φ, a fixed phase difference between incoming and reflected wave. The formula, which is independent of y, represents a standing wave for any fixed x-value but a propagating wave for any fixed z-value traveling with a velocity $c/\cos \alpha$ in $+x$-direction. Its amplitude is $2A_0 \sin(k_0 z \sin \alpha)$, which periodically varies with height z due to the sine factor. There are nodal and antinodal planes parallel to the surface where the sine factor becomes zero or unity, respectively. These planes follow one another at a distance or period a, normal to the surface plane:

$$a = \frac{\lambda}{2 \sin \alpha} \qquad (2\text{-}12)$$

At normal incidence, this equation is reduced to

$$A_{int} = 2A_0 \sin(k_0 z) \cos(k_0 ct + \varphi) \qquad (2\text{-}13)$$

This equation describes the most familiar standing wave with a period of $\lambda/2$.

In order to determine the radiation intensity within the wave field, the electric-field strength has to be squared at any point in space. A general formula for the intensity can even be derived [8] by assuming a reflectivity $R < 100\%$ for the substrate. In front of a flat substrate, the intensity I_{int} is given by

$$I_{\text{int}}(\alpha, z) = I_0 \left[1 + R(\alpha) + 2\sqrt{R(\alpha)} \cos(2\pi z/a - \phi(\alpha)) \right] \qquad (2\text{-}14)$$

where I_0 is a measure of the intensity of the *primary* beam, which is supposed to be constant in time and space. The argument of the cosine is the phase difference of the incoming and reflected waves, including two components: a travel distance $2\pi z/a$ and a phase shift ϕ. This phase shift [8] only occurs in the region of total reflection and is determined by

$$\phi(\alpha) = \arccos \left[2(\alpha/\alpha_{\text{crit}})^2 - 1 \right] \qquad (2\text{-}15)$$

It falls from π to 0 if α is changed from 0 to α_{crit}. For $\alpha > \alpha_{\text{crit}}$, ϕ is continuously zero.

The intensity I_{int} given by equation (2-14) is dependent on α and z but independent of x and y. Nodes and antinodes can be characterized by minimum of maximum intensity, respectively:

$$I_{\text{min,max}} = I_0 \left[1 + R(\alpha) \mp 2\sqrt{R(\alpha)} \right] \qquad (2\text{-}16)$$

The highest contrast between minimum and maximum is reached for $R = 100\%$, leading to $I_{\text{min}} = 0$ and $I_{\text{max}} = 4$. This fact can be explained in the following simple way. In the case of total reflection, the amplitude of the incoming wave and the reflected wave are equal and can be subtracted so as to go down to zero or added so as to double in value. Squaring of these extreme values gives the aforementioned results of 0 and 4, respectively, for the intensities.

Within the substrate, the intensity is exponentially decreasing with the depth z according to

$$I_{\text{int}}(\alpha, z) = I_0 \left[1 + R(\alpha) + 2\sqrt{R(\alpha)} \cos \phi \right] \exp(-z/z_n) \qquad (2\text{-}17)$$

where z_n is the penetration depth defined by equation (1-43). Both functions of intensity (2-14) and (2-17) coincide at the surface of the substrate, i.e., for $z = 0$. This continuity is of course a must.

The dependence of the intensity on the height above or the depth below a surface is illustrated in Figure 2-9 for three different angles α. Silicon was chosen as the flat substrate and Mo-$K\alpha$ radiation as the monochromatic X-ray beam. (a) For $\alpha_{\text{crit}} = 0.1°$, nodes and antinodes follow with a distance $a_{\text{crit}} = 20$ nm and the first antinode coincides with the surface (at $z = 0$). With $R_{\text{crit}} = 81.5\%$, antinodes have a 3.6-fold intensity of the primary beam and the nodes have a nearly zero intensity (about 0.01-fold). Below the surface, the

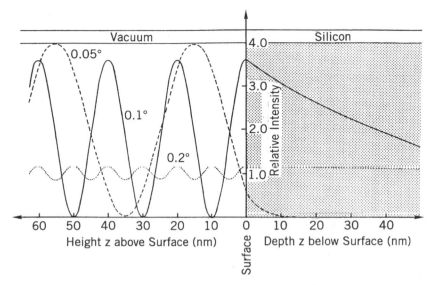

Figure 2-9. Primary intensity above and below a thick Si-flat (from left to right). An X-ray beam of Mo-$K\alpha$ is assumed to strike the flat at glancing angles of $0.05°$ (----), $0.1°$ (——), and $0.2°$ (·····). A standing wave field distinctly arises above the surface for glancing angles at and below the critical angle of total reflection, which is $0.1°$.

intensity is exponentially damped within a depth of some 10 nm. (b) For decreasing angles $\alpha < \alpha_{crit}$, nodes and antinodes are stretched while the first antinode is moving away from the surface. The intensity of the antinodes is increased to a nearly 4-fold value. In the substrate, the intensity is evanescent within only a few nanometers. (c) For increasing angles $\alpha > \alpha_{crit}$, nodes and antinodes are compressed toward the surface. The oscilliations of intensity vanish, and the intensity approaches unity. Below the surface, the intensity reaches a depth of several micrometers instead of only some nanometers.

At the critical angles α_{crit}, the period of standing waves is determined by

$$a_{crit} = \frac{\lambda}{2} \frac{1}{\sqrt{2\delta}} \tag{2-18}$$

This quantity may be of special interest. It is correlated to the minimum penetration depth z_0, defined by the first of equations (1-44). Both quantities a_{crit} and z_0 only differ by the factor 2π. Like the minimum depth z_0, the period a_{crit} is a material constant that is independent of the wavelength or photon energy of the primary beam. Table 2-1 summarizes a_{crit}- and z_0-values calculated according to equations (2-18) and (1-45), respectively.

Table 2-1. Distance a_{crit} of Nodal Planes in Front of Various Substrates and Minimum Penetration Depth z_0

Substrate	a_{crit} (nm)	z_0 (nm)
Plexiglas	26.9	4.3
Glassy carbon	25.6	4.1
Boron nitride	20.3	3.2
Quartz glass	20.3	3.2
Aluminum	18.9	3.0
Silicon	20.0	3.2
Cobalt	10.7	1.7
Nickel	10.4	1.7
Copper	10.7	1.7
Germanium	14.0	2.2
Gallium arsenide	14.0	2.2
Tantalum	8.3	1.3
Platinum	7.3	1.2
Gold	7.7	1.2

2.2.2. Standing Wave Fields Within a Thin Layer

Next to an infinitely thick substrate, a thin homogeneous layer on such a substrate is of particular interest. The paths of X-rays for the two different cases A and B have already been shown in Figure 2-1. Constructive or destructive interference was demonstrated for the two reflected beams propagating in the same direction. They result in Kiessig maxima and minima of the reflectivity appearing for certain angles or *directions* α_k.

Both reflected beams not only interfere with each other, they also jointly interfere with the incoming beam though propagating in a different direction. As demonstrated in Figure 2-10, the beams cross over in a triangular region I in front of the layer and in a trapezoidal region II within the layer. They overlap under an angle $2\alpha_0$ in front of the layer and under an angle $2\alpha_1$ within the layer. This superposition gives rise to standing waves, as was already shown in the foregoing subsection concerning a thick substrate. The standing wave pattern in both regions essentially looks like that of Figure 2-8(b). Nodes and antinodes run parallel to the surface and to the interface. In region III of the underlying substrate, a simple propagating wave appears with amplitudes decreasing with depth. A propagating wave also arises in region IV, where only the two reflected waves overlap. Its amplitude strictly depends on the angle α_0, showing maxima and minima as was demonstrated in Section 2.1.1.

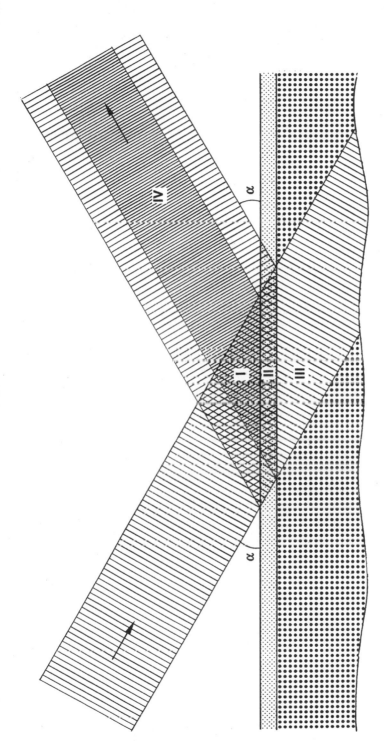

Figure 2-10. Regions of interference in front of and within a thin layer (small dots) deposited on a thick substrate (big dots). Standing waves appear in the triangular region I and in the trapezoidal region II. Regions III and IV show simple propagating waves with a depth dependent amplitude (region III), or an angle dependent amplitude (region IV).

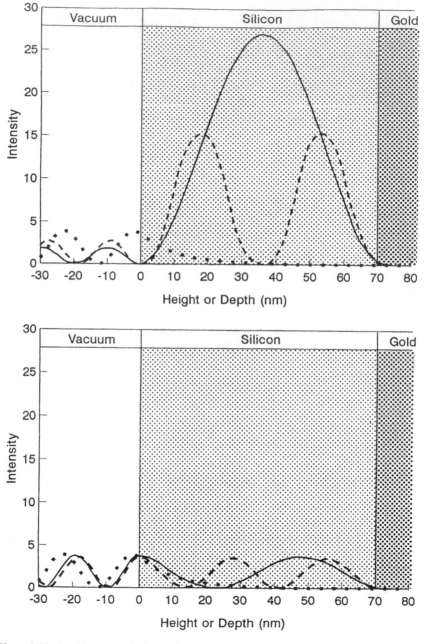

Figure 2-11. Standing waves in front of and within a thin layer deposited on a thick substrate. Case A: silicon on gold. The normalized intensity is plotted vs. the depth normal to the 70 nm Si-layer at various glancing angles. Above: $\alpha = 0.097°$ (▪▪▪▪▪); $0.103°$ (——); $0.111°$ (----). Below: $\alpha = 0.097°$ (▪▪▪▪▪); $0.106°$ (——); $0.117°$ (----). (After de Boer [3].)

As just noted, standing waves are expected to arise in a triangular and trapezoidal region in front of and within the layer, respectively. Of course, the incoming beam has to be monochromatic and plane. In that case, the intensity of the radiation field can be expected to depend only on the height or depth z normal to the layer and not on the wavelength λ.

The intensity of the X-radiation field again results from squaring the amplitudes of the electromagnetic wave at any point in the field. This intensity normalized to the intensity of the primary incoming beam is demonstrated in Figure 2-11 for case A and in Figure 2-12 for case B. These two cases, chosen from refs. 3 and 4, respectively, have already been considered in Section 2.1.1 The interfaces of the layer are indicated by vertical lines. The intensity curves are calculated for five different glancing angles of an incoming Mo-$K\alpha$ beam.

In both cases A and B, nodes and antinodes can be realized in front of and within the thin layer. *In front of the layer*, they are extremely marked at angles below the critical angle of the layer (dotted lines in both Figures 2-11 and 2-12, above and below). Nodal lines with zero intensity follow in height with a periodicity of $\lambda/2\alpha$ according to equation (2-12). The antinodes in their midst have a fourfold intensity, as already observed for a thick substrate. With increasing angle α, the standing wave is generally more and more lacking in contrast. *Within the layer*, the incoming wave is evanescent for angles below the critical angle of the layer (dotted lines in Figures 2-11 and 2-12). Just above this angle, however, standing waves arise with distinct nodes and antinodes. With increasing angle α, more and more periods fit between the two interfaces according to equation (2-12). In the upper parts of both figures, standing waves are shown for angles of the first and second Kiessig minima of reflectivity (full and dashed lines). The lower parts of both figures represent standing waves for the first and second Kiessig maxima of reflectivity determined by equation (2-4). From these figures, it can be observed that an integral number of half periods of a standing wave fit between the interfaces in case of a Kiessig minimum or maximum. This number is even for a minimum and odd for a maximum. The first case may be called *resonance*; the second, *antiresonance*. Resonances within the layer obviously correspond to minima of reflectivity; antiresonances correspond to maxima.

The resonances lead to especially high antinodes lying symmetrically in the layer. In case A, the intensity is still higher (27-fold) than in case B (8-fold). The excessive intensity can be explained by the layer acting as a waveguide [3]. Radiation is swapped back and forth in this waveguide and consequently is reinforced. The extremely high antinode of Figure 2-11 corresponds to the first extremely deep break in the reflectivity curve of Figure 2-2. X-ray waveguide effects are specifically dealt with and described by Zheludeva et al. [9].

Cases A and B mainly differ in the contrast of nodes and antinodes within the layer. This contrast is more distinct in case A (Figure 2-11) than in case

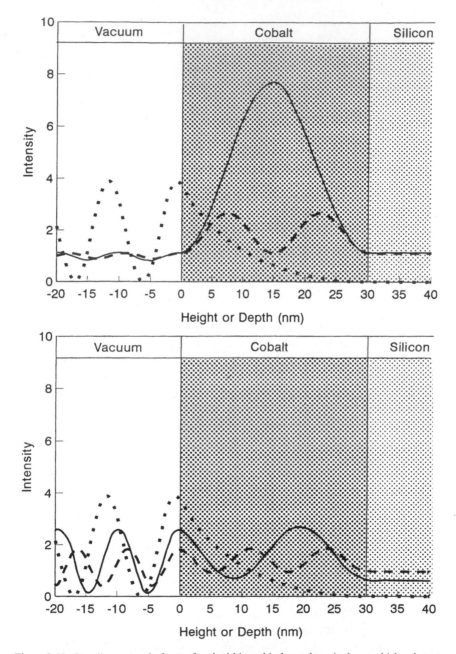

Figure 2-12. Standing waves in front of and within a thin layer deposited on a thick substrate. Case *B*: cobalt on silicon (after ref. 4). The normalized intensity is plotted vs. the depth normal to the surface of the 30 nm Co-layer at various glancing angles: Above: $\alpha = 0.189°$ (▪▪▪▪); $0.203°$ (——): $0.235°$ (----). Below: $\alpha = 0.189°$ (▪▪▪▪); $0.214°$ (——); $0.253°$ (----).

B (Figure 2-12). A further difference can be recognized for the underlying *thick substrate*. In case A the X-ray beam is evanescent in the substrate gold, whereas in case B it deeply penetrates into the substrate silicon.

2.2.3. Standing Wave Fields Within a Multilayer or Crystal

Standing wave fields also arise in multilayers and even in crystals. They can attain a high degree of symmetry especially in crystals but also in *periodic multilayers*. In Figure 2-4, a periodic multilayer was demonstrated with the geometric pathways of X-rays. Transmitted and reflected beams overlap under an angle $2\alpha_1$ in all reflectors and under $2\alpha_2$ in all spacers. In accord with the aforementioned considerations, standing waves are formed in each layer with nodes and antinodes parallel to the interfaces [10]. For perfect crystals, this phenomenon has long been known and is described by the *dynamical theory of diffraction* [11,12]. It is based on the coherent scattering of X-rays at the individual atoms of the different crystal lattice planes.

In a rough approximation, such a multilayer can be considered a single layer of thickness $D = Nd$ to which the relations of Section 2.1.1 are applicable to a certain degree. If the glancing angle α just exceeds the critical angle α_{01} of the upmost reflector layer, an incoming monochromatic beam will deeply penetrate into the multilayer and produce a standing wave field. With increasing angle α, more and more nodal and antinodal planes will be formed in parallel to its surface. At distinct angles α_k they will exactly fit in the multilayer and give rise to relative maxima or minima of reflectivity introduced earlier as Kiessig maxima or minima. The corresponding glancing angles α_k follow equation (2-4).

The exceptional case of Bragg reflection becomes apparent from Figure 2-13. If the antinodes or nodes coincide with all reflectors (or crystal planes), the upmost degree of symmetry will be reached. The periodicity of the standing wave is then equal to that of the multilayer. The antinodes can be situated either in the spacers (Figure 2-13a) or in the reflectors (Figure 2-13c). This shift happens due to the phase jump of π occuring if reflection changes from the optically denser spacer to the optically thinner reflector [3,10]. Figure 2-13b illustrates Bragg reflection when the antinodes lie on top of the reflectors and the nodes on top of the spacers. Bragg reflection appears for the Nth Kiessig maximum when k corresponds to N, the number of bilayers, or an m-fold of this value ($k = Nm$). In this case Bragg's law is valid for glancing angles α_m, determined by equations (2-8) or (2-9). These angles are distinctly above the critical angle α_{01} of total reflection but small enough for the applied sine approximation. These α_m-values represent the Bragg maxima of reflectivity for a periodic stratified medium.

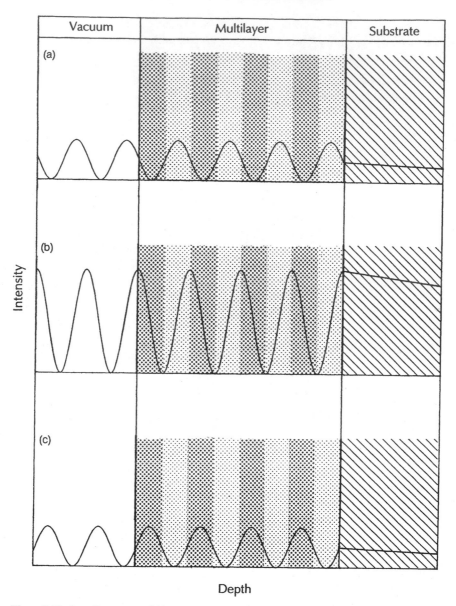

Figure 2-13. Standing waves within a multilayer consisting of a stack of four bilayers for the case of Bragg reflection. The reflectors are dotted darker; the spacers, dotted lighter. Case (a) represents an angle just below the Bragg angle, which is exactly reached in case (b), while case (c) represents an angle just above the Bragg angle. In all three cases, the wave pattern has the periodicity of the multilayer itself, but the antinodes jump from the spacers in case (a) to the reflectors in case (c). The Bragg angle with the highest reflectivity is reached in case (b).

2.3. INTENSITY OF FLUORESCENCE SIGNALS

As already mentioned in Section 1.3.1, an X-ray photon with a sufficient energy can be absorbed by the photoelectric process. An inner electron of the absorbing atom is ejected during this process. The instanteneous rearrangement of electrons leads to the emission of X-ray photons called X-ray fluorescence (XRF). The emitted photons give rise to the "characteristic" spectrum representing a fingerprint of atoms in a corresponding sample. The element composition can be estimated from the relative intensities of the individual characteristic peaks or lines. This is the basis of XRF analysis.

Excitation to fluorescene is conventionally done by a centimeter-wide primary X-ray beam representing a propagating wave. The intensity of the wave field is assumed to be locally constant *in vacuo* and to be exponentially decreasing in solids of micrometer or millimeter thickness. In TXRF, however, the primary beam appears as a standing wave field with locally dependent oscillations or as an evanescent wave field. The atoms in these fields are excited to fluorescence with a probability directly proportional to the wave-field intensity. Consequently, the fluorescence signal reflects the intensity of the standing or evanescent wave field in the sample and additionally indicates the elemental composition. On the one hand, an internal standard can be used to compensate for the inhomogeneity of the primary field. On the other hand, the fluorescence signal can indicate the varying field intensity, and its dependence on the glancing angle can even be used for depth profiling of layered structures.

In the following subsections, the fluorescence signals of various systems will be calculated and discussed: signals of a thick and flat substrate, of a residue left on a substrate, and of near-surface layers deposited *on* or buried *in* a substrate. The primary beam shall be monoenergetic, with an intensity I_0 hitting the substrate under a glacing angle α.

2.3.1. Infinitely Thick and Flat Substrates

The most simple case is that of an infinitely thick homogeneous substrate. It is presupposed to be optically flat in order to ensure total reflection of X-rays. The primary intensity I_{int} within such a substrate is already known from equation (2-17). In a depth z of the substrate, the primary beam may induce a fluorescence signal of a constituent element. Naturally, its intensity is proportional to the primary intensity at this depth. The sum of all signals emitted from the entire thick substrate is obtained by integrating equation (2-17) from zero to infinity. If self-absorption is neglected, the total intensity I_B

of the substrate will be

$$I_B(\alpha) = I_n \left[1 + R(\alpha) + 2\sqrt{R(\alpha)}\cos\phi \right] z_n(\alpha) \qquad (2\text{-}19)$$

where $R(\alpha)$ is the reflectivity given by equation (1-41); $z_n(\alpha)$ is the penetration depth determined by equation (1-43); ϕ is the phase shift defined by equation (2-15); and I_n is a norm or reference for the fluorescence intensity registrated by a detector, which is proportional to the primary beam intensity I_0.

Equation (2-19) is not very transparent but can be transformed into a more comprehensible approximation, as will be shown in Section 2.4.1:

$$I_B(\alpha) = I_n C \left[1 - R(\alpha) \right] \alpha \qquad (2\text{-}20)$$

where C is a quantity of the substrate mainly determined by $1/\beta$ or $1/[(\mu/\rho)\rho]$, respectively. This formula can easily be interpreted as follows [13,14].

If the glancing angle α is far above the critical angle α_{crit} of total reflection, the primary X-ray beam deeply penetrates into the substrate. A thick layer is passed through with a thickness proportional to α or to the sine of α if larger glancing angles are considered. The fluorescence signal comes from this layer. Since the primary beam is nearly not reflected ($R \approx 0$), the signal is only dependent on α, i.e., directly proportional to α. In the region of total reflection, however, the primary beam is evanescent in the substrate, as was demonstrated in Figure 2-9. A major part of the primary beam is reflected ($R \gg 0$), and only the remainder $(1 - R)$ is decisive for fluorescence. The quantity $(1 - R)\alpha$ in equation (2-20) is called the energy transfer, defining that portion of the impinging energy that penetrates into the substrate.

Equation (2-20) is demonstrated by Figure 2-14 for a Mo-$K\alpha$ beam striking a thick and flat Si-substrate. In general, the signal intensity linearly decreases as glancing angles α lessen. At the critical angle of total reflection, however, a steplike decrease of the signal intensity can be noticed. This decrease is on the order of 10^{-4}.

Infinitely thick and flat substrates are usually applied as sample carriers for TXRF analyses. Polished plates of silicon or quartz glass, for example, are highly suitable. They give rise to a very low blank or background spectrum consisting of fluorescence peaks, e.g., Si-$K\alpha$, and a scatter part with tube peaks and a continuum. The intensity of the *fluorescence* and of the *scatter* signals are equally dependent on the energy transfer as described by the intensity I_B [13,14]. The index B was chosen in order to signify blank or background. At total reflection, the angle α is small, $R(\alpha)$ is nearly 1 and the intensity I_B becomes extremely small. For light substrates like silicon, quartz glass, and also Plexiglas, the blank or spectral background of TXRF is 6 orders of

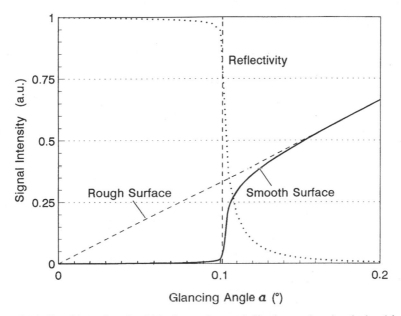

Figure 2-14. Signal intensity of a thick, flat, and smooth Si-substrate (———), calculated for an impinging Mo-$K\alpha$ beam. In addition, the reflectivity R (.....) is shown, dependent on the glancing angle α. Below $\alpha_{crit} = 0.102°$, total reflection occurs with a steplike increase in reflectivity and a steplike decrease of the signal intensity. The oblique dashed line represents the intensity from a rough Si-substrate.

magnitude smaller than that of conventional XRF. This is the reason for the vary low detection limits of TXRF.

2.3.2. Granular Residues on a Substrate

In addition to an infinitely thick and flat substrate, an analyte sample is of particular interest. It may be deposited on such a substrate used as a sample carrier and may be a mineral residue of a solution or suspension after drying. But it may also be a thin section of an organic tissue, or a thin-film-like residue of a metallic smear [15]. For analysis, the analyte is positioned within the standing wave field in front of the substrate. It should not disturb this primary wave field significantly. For example, it should be so rough that a total reflection of the primary beam does not happen at the analyte itself but only at the substrate. As already mentioned, the analyte is excited to fluorescence and the emitted fluorescence signal is proportional to the primary-field intensity. The intensity of this field with nodes and antinodes is given by equation (2-14) and illustrated by Figure 2-9.

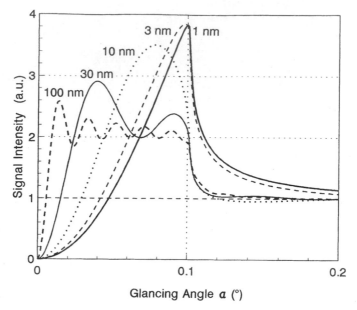

Figure 2-15. Fluorescence signal of small particles or thin films deposited on a silicon substrate used as sample carrier. The intensity was calculated for particles, thin films, or sections of different thickness but equal mass of the analyte, and plotted against the glancing angle α. A Mo-$K\alpha$ beam was assumed for excitation. Particles or films more than 100 nm thick show double intensity below the critical angle of 0.1°.

The analyte may consist of one or several particles of the same grain size s, deposited side by side, or may be a thin section or film with thickness s. In this case, the totally emitted intensity of the analyte is obtained by integrating equation (2-14) between $z = 0$ and $z = s$. The corresponding result is shown in Figure 2-15 for different grain size or thickness but equal mass of the analyte. The fluorescence intensity is independent of the glancing angle α if this angle is far beyond the critical angle of total reflection. In the angular range of total reflection, however, strong oscillations occur for granular or thin-film-like-residues smaller than about 100 nm. These oscillations develop since only a few antinodes and nodes penetrate the particles, thin films, or sections. They diminish with increasing grain size or thickness and finally approach a constant value [14,16]. It results from the approximation

$$I_x(\alpha) = I_n m_x [1 + R(\alpha)] \qquad (2\text{-}21)$$

where m_x is the mass of the analyte element x on top of the substrate. The constant value of the intensity is $(1 + R)$-fold, or nearly double, the amount observed for angles beyond the total-reflection zone. The doubling can easily be explained by the fact that particles are excited equally by the incoming beam *and* the reflected beam. Thus, if the glancing angle falls short of the critical angle, the fluorescence intensity will step up to the double value.

For quantitative analyses, it is important to get an intensity that is independent of particle size or film thickness and also of glancing angle. This ideal behavior is realized for granular or thin-film-like samples of about several 100 nm. But also several particles of various thicknesses may be applied, as frequently happens in practice. A broad size distribution can average over the annoying oscillations [17]. On the other hand, the particles have to be small enough in order to avoid absorption effects [15]. In any case, the need for internal standardization is obvious. A standard may be added to the particles and distributed homogeneously over the height of the particles. In particular, the standard must not be enriched in the nodes or antinodes of the primary field but should be uniformly mixed with the sample. In case of a fairly homogeneous distribution, the intensity ratio of analyte and internal standard is constant, i.e., independent of the particle size or thickness and also of the glancing angle α.

2.3.3. Buried Layers in a Substrate

After our consideration of an analyte placed *on* a substrate, we shall now consider an analyte that is included *in* a substrate. This analyte can be a layer containing either impurities or minor but essential constituents within the substrate. The layer can be produced by evaporation or implantation. Such layers in wafers are called buried layers. If they are localized directly below the surface of the substrate, they may be called near-surface layers. But they can also represent a deeper lying interface. In any case, they are assumed to contain the decisive constituents in a low concentration. Consequently, the refractive index of the substrate can be considered unchanged and the layer itself can be assumed to be nonreflecting. Furthermore, the concentration is assumed to be constant within the layer, representing a rectangular profile with a thickness d.

The fluorescence intensity emitted from this layer can be calculated by integration of equation (2-17) between the boundaries of the layer. If the layer is between the depth z and $z + d$ and if absorption can be neglected, the result will be given by

$$I_{BL}(\alpha,z) = I_n c_A C \left[1 - R(\alpha)\right] \frac{\alpha}{d} \exp\left(-\frac{z}{z_n}\right) \left[1 - \exp\left(-\frac{d}{z_n}\right)\right] \quad (2\text{-}22)$$

in accord with Schwenke and colleagues [13,14]. The index BL signifies the buried layer; c_A is the area-related mass (g/cm^2) or area density (atoms or

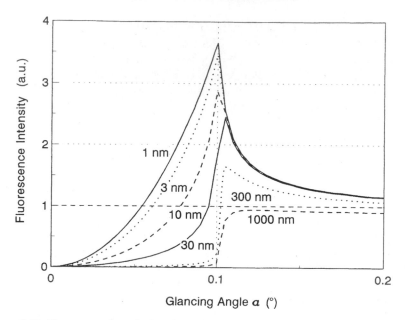

Figure 2-16. Fluorescence intensity from layers buried in a thick substrate. The intensity dependent on the glancing angle was calculated for layers of different thickness but with a constant area density of the analyte. Silicon was assumed as the substrate; Mo-$K\alpha$ X-rays, as the primary beam. Total reflection occurs in the region below 0.1°. Without total reflection, the dashed horizontal line would be valid in all.

ions/cm^2) of the given element; C is the quantity defined in equation (2-20); R is the reflectivity of the substrate; and z_n is the penetration depth of the primary X-ray beam normal to the surface. For ultrathin layers directly below the surface, i.e., for $z = 0$ and $d = 0$, equation (2-22) leads to

$$I_{BL}(\alpha) = I_n c_A C \left[1 - R(\alpha)\right] \frac{\alpha}{z_n} \tag{2-23}$$

The dependence of the fluorescence intensity on the glancing angle α is demonstrated in Figure 2-16 for buried layers of different thickness d but equal area density c_A. For the extremely thin layer of 1 nm, the curve is identical to that of Figure 2-15 for the same thickness. This agreement is caused by a continuous transition of the primary intensity into the substrate even at total reflection.

Thus a single function can be expected for ultrathin layers *above* or *below* the surface of a substrate. It can be derived from equation (2-23) by use of equations (2-19), (2-20), and (2-15) and can be approximated by

$$I_{BL}(\alpha) \simeq I_n c_A \, 4\sqrt{R(\alpha)} \left(\frac{\alpha}{\alpha_{crit}}\right)^2 \tag{2-24}$$

in the region of total reflection $\alpha \leqslant \alpha_{crit}$, and by

$$I_{BL}(\alpha) \simeq I_n c_A \left[1 + \sqrt{R(\alpha)}\right]^2 \tag{2-25}$$

in the region beyond total reflection. The first approximation represents a parabola with a maximum intensity of $4\sqrt{R}$ at the critical angle. The second equation describes an asymptotic decrease of the curve to a constant value at larger angles, as illustrated by Figure 2-16.

For layers of some 100 nm, the thickness d far exceeds the penetration depth z_n. In this case the peaks at the critical angles are reduced and the curves of Figure 2-16 get an S-like shape. Equation (2-22) can be approximated by

$$I_{BL}(\alpha) \simeq I_n c_A C \left[1 - R(\alpha)\right] \frac{\alpha}{d} \tag{2-26}$$

with symbols as defined above.

Finally, an infinitely thick layer may be considered. This layer may be thought of as a substrate in which the analyte element is homogeneously distributed. Equation (2-26) leads to the asymptotic expression

$$I_{BL}(\alpha) \simeq I_n c_v C \left[1 - R(\alpha)\right] \alpha \tag{2-27}$$

where c_v is the concentration of the analyte in the total volume of the substrate given by the limit of c_A/d. Expression (2-27) is already known from equation (2-20), but it additionally contains the quantity c_v as the volume concentration.

2.3.4. Reflecting Layers on Substrates

The granular residues and buried layers treated so far were both assumed to be nonreflecting. Their fluorescence signal was calculated by the rather simple and transparent equations (2-21)–(2-27). By way of contrast, reflecting layers shall now be considered. They may be plated on top of thick substrates in order to be used as conductive or nonconductive layers in wafer technology, as protective or decorative coatings, etc. First of all, let us consider single reflecting layers. They are assumed to be flat, smooth, and equally thick over

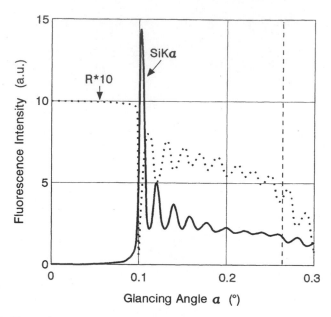

Figure 2-17. Fluorescence intensity from a thin layer deposited on a thick substrate. Case *A* of Section 2.1.1 was assumed for the calculation: a 70 nm Si-layer on a Au-substrate irradiated by a Mo-$K\alpha$ beam. The intensity of Si-$K\alpha$ (——) was determined in dependence on the glancing angle α. From Figure 2-2, the reflectivity of the layered substrate is shown again (.....). It can be verified that Kiessig maxima of reflectivity correspond to minima of fluorescence, and vice versa. (After de Boer [3].)

the entire surface of a substrate. The thickness may range from about 0.2 nm for a monoatomic layer up to some 100 nm.

The X-ray-induced fluorescence of such layers can be calculated by means of fundamental parameters. In a first step, the primary intensity within the layer has to be calculated as a function of the depth, as demonstrated for example in Figure 2-11 or 2-12. In a second step, the excitation to fluorescence, the absorption and enhancement must be incorporated into the calculation. Finally, the fluorescence intensity values have to be integrated over the layer thickness. The mathematics is somewhat complex and will be presented later on in Section 2.4. To start with, the fluorescence signal shall be shown only for a few examples. In particular, the angular dependence of this signal shall be represented, as has already been done for nonreflecting layers.

In Figures 2-17 and 2-18, the fluorescence intensity emitted from a single layer is shown for the two cases *A* and *B* that were already considered in Sections 2.1.1 and 2.2.2. Case *A* represents a 70 nm Si-layer deposited on a Au-substrate; case *B*, a 30 nm Co-layer on a Si-substrate. A Mo-tube was chosen

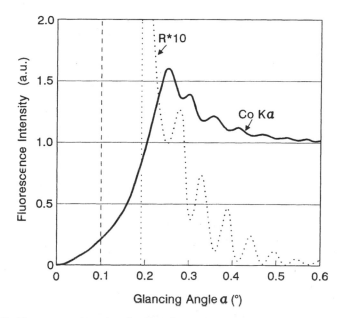

Figure 2-18. Fluorescence intensity of a thin reflecting layer deposited on a thick substrate. Case B of Section 2.1.1 was considered: a 30 nm Co-layer on Si, irradiated by Mo-$K\alpha$ X-rays. The Co-$K\alpha$ intensity (——) is dependent on the glancing angle α and connected with the reflectivity of the layered sample (.....). (After de Boer and van den Hoogenhof [4].)

for excitation. The measured signal of the Si-$K\alpha$ or Co-$K\alpha$ radiation is demonstrated in angular dependence [3,4]. Below the critical angle of the single layer, the intensity is low but increases strongly at this angle, especially in case A. Beyond the critical angle of the layer, the intensity shows oscillations with maxima and minima but then approaches a constant value. As Figures 2-17 and 2-18 demonstrate, the maxima and minima of the fluorescence intensity correspond to the minima and maxima of the reflectivity of the layered substrates. These extrema are named Kiessig minima and maxima, as already mentioned in Section 2.1.1. A low reflectivity of course means a high energy flow into the layer and consequently a strong excitation to fluorescence. For high reflectivity, the relations are vice versa. The steep Si-$K\alpha$ peak in Figure 2-17 corresponds to the abrupt drop in reflectivity of Figure 2-2. It also corresponds to the excessive primary intensity demonstrated in Figure 2-11 as an extreme case of resonance.

The relationships of course depend on the thickness of the deposited layers. For the foregoing example of case B, the influence of the layer thickness is demonstrated in Figure 2-19. The thickness of Co-layers varies between 1 and 200 nm, leading to a broad scale of intensity values for the Co-$K\alpha$ radiation

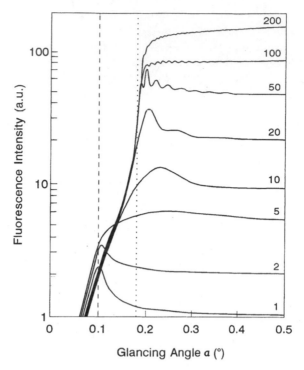

Figure 2-19. Angular dependence of the fluorescence radiation emitted from a Co-layered Si substrate. The Co-$K\alpha$ intensity is plotted semilogarithmically for layers of different thickness (in nm). The maxima for the ultrathin Co-layers are localized at the critical angle of Si (dashed vertical line). They are shifted to the critical angle of Co (dotted vertical line) if the layer is more than 10 nm thick. (After Schwenke et al. [16].)

[16]. Consequently, a semilogarithmic plot is chosen to provide a better general view.

For ultrathin layers, the curves are similar to those for nonreflecting layers. Since the radiation penetrates the nanometer-thin Co-layers, total reflection does not occur at the Co-layer but only at the Si-substrate. For that reason, the intensity maximum appears at the critical angle of Si. If the layer exceeds a thickness of 10 nm, the maximum, however, shifts to the critical angle of Co. The range of intermediate thickness between 10 and 100 nm is characterized by oscillations of intensity beyond the Co-critical angle. They are already known from Figures 2-17 and 2-18 and defined as Kiessig maxima and minima. Their angular period as well as their contrast decreases with thickness. Finally, the curves smooth out and show the shape of an infinitely thick substrate. On that account, the 200 nm thick layer gives a curve like that shown in Figure 2-16.

The last example pertains to case *B*. The first maximum shifts from the critical angle of the substrate to that of the layer if its thickness exceeds 10 nm or so. In particular, the maximum shifts to a greater angle. Such a behavior also occurs for examples of case *A*, but there the maximum shifts to a smaller angle [18].

2.3.5. Periodic Multilayers and Crystals

Single reflecting layers will present the characteristic features of TXRF if the glancing angle of the primary beam is varied at grazing incidence. In principle, multilayers show a similar angular dependence of the fluorescence intensity with typical Kiessig maxima and minima. The intensity can be calculated by integration over the standing wave field within the layers. The calculation is carried out by computers using a fundamental parameter approach, as will be described in Section 2.4. It can be simplified due to the small glancing angles considered.

A further phenomenon will occur if a *periodic* multilayer is taken into consideration [10]. The same effect happens for natural or synthetic crystals, as they are akin to periodic multilayers already considered in Section 2.1.2. Figure 2-20 [19] demonstrates this effect for a multilayer already chosen in Figure 2-6. It may consist of 15 bilayers of 1.9 nm Pt and 0.2 nm Co, with a period of about 2.1 nm. The fluorescence intensity in Figure 2-20 strongly increases around the critical angle of Pt at 0.6°. It further shows maxima at angles of minimum reflectivity and minima at angles of maximum reflectivity. These Kiessig oscillations, however, are not very distinct in comparison to the following Bragg maxima and minima of intensity.

The first Bragg maximum of reflectivity ($m = 1$) occurs at a glancing angle of about 2.2°. In the ascending flank of this peak, a maximum of the Co-$K\alpha$ intensity appears whereas a minimum of the Pt-$L\alpha$ intensity occurs. Recall from Figure 2-13 that the antinodes are in the Co-layers while the nodes are in the Pt-layers. If the glancing angle just exceeds the Bragg angle, the antinodes are switched into the Pt-layers and the nodes into the Co-layers due to the phase shift of π. Consequently, the Pt-$L\alpha$ intensity shows a maximum in the descending flank of the reflection peak whereas the Co-$K\alpha$ intensity shows a minimum. The effect is quite distinct for the first order and will recur with smaller contrast at higher orders m. In any case, the intensity maxima of the lighter spacer appear at angles just below those of the heavier reflector.

The phenomenon is also known for crystal lattices, with the following two distinctions. (i) Since there is no spacer element but possibly more than one reflector element, e.g., for NaCl, the intensity minima of these elements always

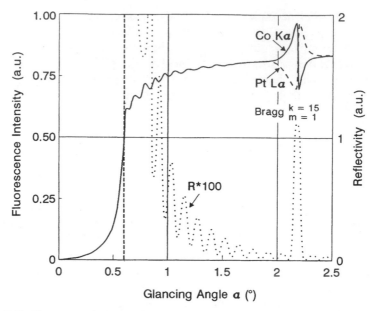

Figure 2-20. Fluorescence intensity of a periodic multilayer plotted against the glancing angle α. The multilayer of Section 2.1.2 was used for calculation, consisting of 15 bilayers of Pt and Co ($d = 2.1$ nm). A Cu-$K\alpha$ beam was assumed for excitation. The reflectivity of the multilayer is represented by the dotted curve. (After Huang and Parrish [7] and de Boer and van den Hoogenhof [19].)

lie on the upgoing flank of the reflection peak, the maxima on the downgoing flank. (ii) Since the lattice distance of anorganic crystals is mostly between 0.15 and 0.3 nm, the Bragg angle is on the order of $10°$ for wavelengths of about 0.1 nm. For that reason, Bragg reflection does not appear under grazing incidence but only under steeper incidence.

2.4. FORMALISM FOR INTENSITY CALCULATIONS

The phenomena qualitatively described in the previous section can rigorously be calculated from theory. The *primary* beam can be described by a plane incoming and reflected wave leading to a standing wave field. Its intensity can be calculated from the optical theory of wave propagation in layers with flat interfaces, especially by the Fresnel relations. The *primary intensity* is a function of the glancing angle α_0 at grazing incidence and is further dependent on the depth z normal to the layer surface. Besides the primary intensity, the intensity ratio of the reflected wave and the incoming wave can be calculated. This ratio is defined as *reflectivity* and can directly be checked by reflectivity measurements. All calculations can be based either on a recursive formalism described by Parratt [20] or on an equivalent matrix formalism first proposed by Abelès [21] and extensively described by Born and Wolf [1] and Król et al. [22]. The latter is the more elegant method and is preferred here.

The *fluorescence intensity* can be determined by a fundamental parameter approach of XRF. This intensity can be calculated as a function of the glancing angle and be checked by fluorescence measurements. The theory has been described and applied by several authors [3, 18, 19, 23–27] and has been proven valid by many experiments [4, 7, 18, 28–31]. Even the surface and interface roughness has been taken into account [26, 32–34].

In the following subsections, the reflectivity as well as the primary and fluorescence intensities are calculated for three different media already distinguished earlier: (i) a thick and flat substrate; (ii) a thin homogeneous layer on a substrate; and (iii) a stratified medium of several layers. The derivations are carried out mainly following papers by Gutschke [18], de Boer and van den Hoogenhof [19], and Weisbrod et al. [28].

2.4.1. A Thick and Flat Substrate

An infinitively thick and flat substrate may be present, and the z-direction shall be assumed normal to the surface (as shown in Figure 2-21). A plane wave may hit the surface under a glancing angle α_0, may be reflected under an equal angle

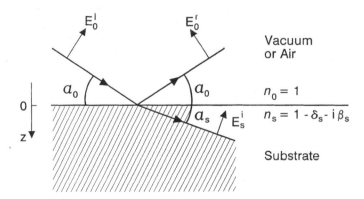

Figure 2-21. Incoming, reflected, and refracted beams at the interface between vacuum or air (above) and a thick substrate (below): n_0 and n_s are the refractive indices; z is the direction normal to the surface; and E is the electric-field vector representing a π-polarization.

α_0, and transmitted under a refraction angle α_s. It is given by

$$\alpha_s = \sqrt{\alpha_0^2 - 2\delta_s - 2i\beta_s} \qquad (2\text{-}28)$$

where δ_s and β_s are the real and complex parts of the refractive index of the substrate, respectively. The real and imaginary components of α_s can be derived in accord with equations (1-36) and (1-37), respectively.

The radiation intensity is determined by the amplitude of the electric-field vector. At steeper incidence, π- and σ-polarization have to be distinguished (the field vector is within the plane of incidence or perpendicular to this plane, respectively). At grazing incidence, however, the polarization does not have to be considered. Only the length or amplitude of the electric-field vector is relevant for fluorescence. Consequently, only this component must be taken into account, but an averaging over time should be carried out.

There are different waves with different amplitudes. The incident wave amplitude E_0^i may be normalized to 1, and the reflected wave amplitude may be denoted by E_0^r. The refracted wave amplitude, which is the incident wave amplitude in the substrate, may be called E_s^i. Within the substrate, there is no reflected wave, so that a hypothetical amplitude E_s^r can be set to zero. These four amplitudes are connected by the so-called Born or transfer matrix according to

$$\begin{bmatrix} 1 \\ E_0^r \end{bmatrix} = \begin{bmatrix} m_1 & m_2 \\ m_3 & m_4 \end{bmatrix} \cdot \begin{bmatrix} E_s^i \\ 0 \end{bmatrix} \qquad (2\text{-}29)$$

The individual components of this 2×2 matrix are given by

$$m_1 = m_4 = \frac{\alpha_0 + \alpha_s}{2\alpha_0} \qquad (2\text{-}29a)$$

$$m_2 = m_3 = \frac{\alpha_0 - \alpha_s}{2\alpha_0} \qquad (2\text{-}29b)$$

The primary intensity I_{int} is defined as the square of the modulus of the relevant amplitude. Since the incoming and reflected beams interfere above the surface, the intensity I_{int} at height z for a glancing angle α_0 can be written as a sum of the downgoing and upgoing plane waves:

$$I_{int}(\alpha_0, z) = I_0 |\exp(-ik_0\alpha_0 z) + E_0^r \exp(ik_0\alpha_0 z - \phi)|^2 \qquad (2\text{-}30)$$

where k_0 is $2\pi/\lambda$; ϕ is a phase shift already defined by equation (2-15); and I_0 is a measure for the incoming beam intensity. Squaring leads to

$$I_{int}(\alpha_0, z) = I_0 \left[1 + |E_0^r|^2 + 2|E_0^r| \cos(2k_0 \alpha_0 z - \phi) \right] \qquad (2\text{-}31)$$

Since $|E_0^r|^2$ is defined as reflectivity R, this equation is equivalent to expression (2-14).

The amplitude E_0^r is obtained from equation (2-29) as the ratio of m_3 and m_1. The squared value of the modulus yields the reflectivity itself:

$$R(\alpha_0) = \left| \frac{\alpha_0 - \alpha_s}{\alpha_0 + \alpha_s} \right|^2 \qquad (2\text{-}32)$$

which corresponds to equation (1-40).

Within the substrate, the primary intensity is determined only by the refracted beam with the amplitude E_s^i.

$$I_{int}(\alpha_0, z) = I_0 |E_s^i \exp(i k_0 \alpha_s z)|^2 \qquad (2\text{-}33)$$

The amplitude E_s^i obtained from equation (2-29) is given by $1/m_1$. Inserting the respective value of m_1 leads to

$$I_{int}(\alpha_0, z) = I_0 \left| \frac{2\alpha_0}{\alpha_0 + \alpha_s} \right|^2 \exp(-z/z_n) \qquad (2\text{-}34)$$

where z_n means the penetration depth defined by equation (1-43).

The quadratic term is the transmissivity. By means of the reflectivity R defined in equation (1-40), this quantity can easily be shown to equal $(1 - R)\alpha_0/\alpha_s'$. Replacing α_s' by α_s'' in accord with equation (1-38) and inserting z_n defined by equation (1-43) leads to

$$I_{int}(\alpha_0, z) = I_0 C (1 - R) \frac{\alpha_0}{z_n} \exp(-z/z_n) \qquad (2\text{-}35)$$

where C is the quantity already used in equation (2-20) and which is equal to $(\lambda/4\pi)/\beta_s$. Equation (2-35) is equivalent to equation (2-17) in Section 2.2.1. It is the basis for the fluorescence intensities of different buried layers that were calculated by integration and led to the approximate equations (2-22)–(2-27).

A more detailed calculation of the X-ray fluorescence intensity can be based on a fundamental parameter approach. It may be provided that the takeoff angle for the detector is 90° and that the detector area determines the area of observation. If an element x with a mass fraction c_x is homogeneously distributed in a thick and flat substrate, the fluorescence intensity can be approximated by

$$I_x(\alpha_0) = I_0 c_x S_{x,E_0} \frac{1-R}{(\mu/\rho)_{s,E_0}/\alpha_0 + (\mu/\rho)_{s,E}} \varepsilon_{det} T_{air} \qquad (2\text{-}36)$$

This equation substitutes for equation (2-27). Here I_0 is the constant intensity of the incident plane wave or primary beam; S_{x,E_0} is a sensitivity value of element x at photon energy E_0 of the monochromatic primary beam; $(\mu/\rho)_{s,E_0}$ and $(\mu/\rho)_{0,E}$ are the mass attenuation coefficients of the substrate at photon energy E_0 of the primary beam and at photon energy E of the detected element peak, respectively; α_0 is the glancing angle of incidence, which is assumed to be small ($< 5°$), so that the actual sine dependence can be ignored; ε_{det} is the efficiency of the detector for X-ray photons of energy E; and T_{air} is their transmission by air on the beam path between the sample and the detector. The sensitivity itself can be further deduced from

$$S_{x,E_0} = g_x \omega_x f_x (\tau/\rho)_{x,E_0} \qquad (2\text{-}37)$$

where g_x is the relative emission rate of the element peak in its series; ω_x is the fluorescence yield of the element x; f_x is its jump factor at the relevant absorption edge and $(\tau/\rho)_{x,E_0}$ is the photoelectric mass-absorption coefficient at the primary photon energy E_0. This product of fundamental parameters is independent of instrumentation or sample matrix and can be calculated for each element x. Different sets of tables can be used for this calculation [35–37].

The foregoing derivation includes the absorption of the primary radiation as well as the fluorescence radiation. Secondary fluorescence or enhancement is not taken into account because it is a second-order process. It is rather complex and may be neglected in the case of grazing incidence (but ought to be considered in principle [3,28]). The angular divergence of the instrument, however, can easily be taken into consideration. For that purpose, equation (2-34) has to be convoluted by a triangle function of α_0 with the width of the aperture.

The intensity dependence on the energy transfer $(1-R)\alpha_0$ was derived earlier in simple terms in Section 2.3.1. The matrix and fundamental parameter approaches have been introduced in the present section in order to provide a more detailed description. Both formalisms can easily be expanded and applied to single- and multiple-layer systems.

2.4.2. A Thin Homogeneous Layer on a Substrate

A single layer with a thickness d_1 may be deposited on a thick and flat substrate, as demonstrated in Figure 2-22. Here $n_0, n_1,$ and n_s shall denote the

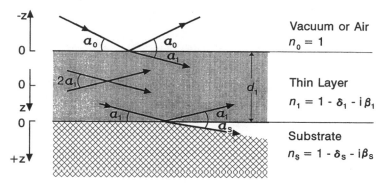

Figure 2-22. Incoming, reflected, and refracted beams above, within, and below a thin layer, respectively, deposited on a thick substrate: n_0, n_1 and n_s are the refractive indices. At any point above and within the layer, there are two beams interfering with one another at an angle $2\alpha_0$ or $2\alpha_1$, respectively. In the substrate, there is only one refracted beam penetrating at an angle α_s.

complex refractive indices of the vacuum, the layer, and the substrate, respectively, dependent on the relevant quantities δ and β. A wide X-ray beam represented by a plane wave may be directed to this simple stratified structure at an angle α_0. In this case, two beams are imaginable at any point above and within the layer: one incoming and one reflected beam. *In vacuo*, they interfere at an angle $2\alpha_0$ and in the layer they overlap at an angle $2\alpha_1$. This is demonstrated for one point at the vacuum–layer interface, for one point at the layer–substrate interface, and for one point in the middle of the layer. In the substrate, however, there are only refracted beams at any point, penetrating at an angle α_s. The angles are determined by

$$\alpha_1 = \sqrt{\alpha_0^2 - 2\delta_1 - 2i\beta_1} \qquad (2\text{-}38a)$$

and

$$\alpha_s = \sqrt{\alpha_0^2 - 2\delta_s - 2i\beta_s} \qquad (2\text{-}38b)$$

respectively.

The radiation intensity of the wave field called primary intensity can easily be calculated by the matrix formalism. For simplification, the origin of the vertical z-axis is defined for the three media separately. For the vacuum and the substrate it is placed at the interfaces; for the layer it is positioned right in the middle of it. Under this condition, the amplitudes of four different electric-field vectors E_0^r, E_1^i, E_1^r, and E_s^i are connected by two transfer matrices

$M_{0,1}$ and $M_{1,s}$ according to

$$\begin{bmatrix} 1 \\ E_0^r \end{bmatrix} = M_{0,1} \begin{bmatrix} E_1^i \\ E_1^r \end{bmatrix} \tag{2-39a}$$

and

$$\begin{bmatrix} E_1^i \\ E_1^r \end{bmatrix} = M_{1,s} \begin{bmatrix} E_s^i \\ 0 \end{bmatrix} \tag{2-39b}$$

The individual components of the first matrix $M_{0,1}$ are given by

$$m_{1,1} = \frac{\alpha_0 + \alpha_1}{2\alpha_0} \exp\left(-i\frac{k_0}{2}\alpha_1 d_1 \right)$$

$$m_{2,1} = \frac{\alpha_0 - \alpha_1}{2\alpha_0} \exp\left(+i\frac{k_0}{2}\alpha_1 d_1 \right)$$

$$m_{3,1} = \frac{\alpha_0 - \alpha_1}{2\alpha_0} \exp\left(-i\frac{k_0}{2}\alpha_1 d_1 \right) \tag{2-40a}$$

$$m_{4,1} = \frac{\alpha_0 + \alpha_1}{2\alpha_0} \exp\left(+i\frac{k_0}{2}\alpha_1 d_1 \right)$$

The components of the second matrix $M_{1,s}$ are determined by

$$m_{1,s} = \frac{\alpha_1 + \alpha_s}{2\alpha_1} \exp\left(-i\frac{k_0}{2}\alpha_1 d_1 \right)$$

$$m_{2,s} = \frac{\alpha_1 - \alpha_s}{2\alpha_1} \exp\left(-i\frac{k_0}{2}\alpha_1 d_1 \right)$$

$$m_{3,s} = \frac{\alpha_1 - \alpha_s}{2\alpha_1} \exp\left(+i\frac{k_0}{2}\alpha_1 d_1 \right) \tag{2-40b}$$

$$m_{4,s} = \frac{\alpha_1 + \alpha_s}{2\alpha_1} \exp\left(+i\frac{k_0}{2}\alpha_1 d_1 \right)$$

The four amplitudes E_0^r, E_1^i, E_1^r, and E_s^i, have to be determined by two matrix equations. By combination of both equations (2-39a) and (2-39b), the amplitudes E_0^r, and E_s^i can first be determined from the relation

$$\begin{bmatrix} 1 \\ E_0^r \end{bmatrix} = M_{0,1} \cdot M_{1,s} \cdot \begin{bmatrix} E_s^i \\ 0 \end{bmatrix} \tag{2-41}$$

Afterward, the two other amplitudes E_1^i and E_1^r of the electric field in the layer can be calculated from equation (2-39b). The reflectivity of the layered substrate already given by equation (2-5) follows from the squared modulus of E_0^r.

The primary intensity in the depth z for a glancing angle α_0 can now be determined for the three different media. In vacuo, the primary intensity results from

$$I_{int}(\alpha_0, z) = I_0 \cdot |\exp(-ik_0\alpha_0 z) + E_0^r \exp(ik_0\alpha_0 z)|^2 \qquad (2\text{-}42a)$$

In the layer it is given by

$$I_{int}(\alpha_0, z) = I_0 \cdot |E_1^i \exp(-ik_0\alpha_1 z) + E_1^r \exp(ik_0\alpha_1 z)|^2 \qquad (2\text{-}42b)$$

while z is in the range $(-d_1/2, +d_1/2)$. In the substrate, the primary intensity follows from

$$I_{int}(\alpha_0, z) = I_0 \cdot |E_s^i \exp(ik_0\alpha_s z)|^2 \qquad (2\text{-}42c)$$

Next, the fluorescence intensity has to be derived for an element x that is present in the layer with a mass fraction c_x. It is assumed again that the detector is positioned perpendicular to the surface with a short gap in between them. The solution follows from the fundamental parameter approach after integration over the thickness d_1 of the layer [19]:

$$I_x(\alpha_0) = I_0 c_x S_{x,E_0} \varepsilon_{det} T_{air} \cdot \left\{ |E_1^i|^2 \cdot \frac{1 - \exp(-[(\mu/\rho)_{1,E_0}/\alpha_1 + (\mu/\rho)_{1,E}]\rho_1 d_1)}{[(\mu/\rho)_{1,E_0}/\alpha_1 + (\mu/\rho)_{1,E}]} \right.$$

$$+ |E_1^r|^2 \cdot \frac{1 - \exp(-[-(\mu/\rho)_{1,E_0}/\alpha_1 + (\mu/\rho)_{1,E}]\rho_1 d_1)}{[-(\mu/\rho)_{1,E_0}/\alpha_1 + (\mu/\rho)_{1,E}]}$$

$$+ 2Re\left[E_1^i \cdot E_1^r \cdot \frac{1 - \exp(-[-2ik_0\alpha_1/\rho_1 + (\mu/\rho)_{1,E}]\rho_1 d_1)}{[-2ik_0\alpha_1/\rho_1 + (\mu/\rho)_{1,E}]} \right] \right\} \qquad (2\text{-}43)$$

The different parameters are defined in correspondence to those of equation (2-36); $(\mu/\rho)_{1,E_0}$ is the mass-attenuation coefficient of the layer at photon energy E_0; $(\mu/\rho)_{1,E}$ the similar coefficient at photon energy E corresponding to the element peak; and ρ_1 the density of the layer. Re means the real part of the complex number in brackets. As mentioned in the previous subsection, the angular divergence can be taken into account by a convolution.

Two special cases may be of interest. For a very thin layer, the thickness d_1 approximates zero and the three fractions of equation (2-43) take on the limiting value $\rho_1 d_1$. The fluorescence intensity simply becomes

$$I_x(\alpha_0) = I_0 c_A S_{x,E_0} \varepsilon_{det} T_{air} \cdot |E_1^i + E_1^r|^2 \tag{2-44}$$

where c_A is the area-related mass given by the product $c_x \rho_1 d_1$. The second special case is a single thin layer for which vacuum or air may be regarded as the substrate. In this case, the amplitude E_1^r nearly vanishes since the reflection at the bottom of the layer can be neglected. Equation (2-43) is then reduced to the first term of the sum.

2.4.3.   A Stratified Medium of Several Layers

To be consistent, we now consider a stratified structure of several layers. It may be composed of N layers with a thickness d_ν ($\nu = 1,...,N$) that are each homogeneous and plane parallel. The refractive indices may be n_ν and the densities ρ_ν. The first layer adjoins to the vacuum or air ($\nu = 0$); the last layer, to the substrate ($\nu = N + 1$). The origin of each layer ($z = 0$) is shifted right in the middle of the respective layer.

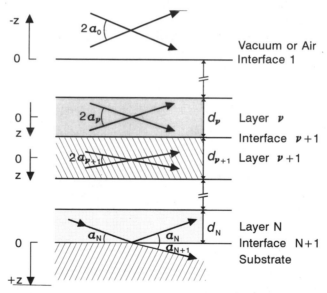

Figure 2-23. Incoming, reflected, and refracted beams in a stratified structure with N layers of a finite thickness d_ν ($\nu = 1,...,N$) deposited on a thick substrate. At any point within the individual layers, one incident beam and one reflected beam interfere with each other at an angle $2\alpha_\nu$. Only in the substrate is there a single incident beam refracted at an angle α_{N+1}. For the vacuum or air and for the substrate, the origin of the z-axis is placed on the interface. But for all layers a separate origin is placed in the middle of the respective layer.

In any layer, there is one definite glancing angle of incidence α_v. It is determined by the first angle α_0 and the quantities δ_v and β_v of the layer, according to

$$\alpha_v = \sqrt{\alpha_0^2 - 2\delta_v - 2i\beta_v} \qquad (2\text{-}45)$$

One incoming and one reflected beam interfere with each other at any point of the vth layer, as shown in Figure 2-23. They overlap under the angle $2\alpha_v$. The amplitudes of the electric fields can be denoted by E_v^i and E_v^r for the incoming and reflected waves, respectively, in the middle of layer v. As in the previous subsection, the amplitudes of two adjacent layers can be connected by a transfer matrix

$$\begin{bmatrix} E_v^i \\ E_v^r \end{bmatrix} = M_{v,v+1} \begin{bmatrix} E_{v+1}^i \\ E_{v+1}^r \end{bmatrix} \qquad (2\text{-}46)$$

with the components

$$m_{1,v+1} = \frac{\alpha_v + \alpha_{v+1}}{2\alpha_v} \exp\left[-i\frac{k_0}{2}(\alpha_v d_v + \alpha_{v+1} d_{v+1}) \right]$$

$$m_{2,v+1} = \frac{\alpha_v - \alpha_{v+1}}{2\alpha_v} \exp\left[-i\frac{k_0}{2}(\alpha_v d_v - \alpha_{v+1} d_{v+1}) \right]$$

$$m_{3,v+1} = \frac{\alpha_v - \alpha_{v+1}}{2\alpha_v} \exp\left[i\frac{k_0}{2}(\alpha_v d_v - \alpha_{v+1} d_{v+1}) \right] \qquad (2\text{-}47)$$

$$m_{4,v+1} = \frac{\alpha_v + \alpha_{v+1}}{2\alpha_v} \exp\left[i\frac{k_0}{2}(\alpha_v d_v + \alpha_{v+1} d_{v+1}) \right]$$

For the vacuum ($v = 0$) and for the substrate ($s = N + 1$), the hypothetical thickness d_0 or d_{N+1}, respectively, has to be set to zero. Due to this agreement, the equations (2-47) are even valid for the components of the first matrix $M_{0,1}$ and the last matrix $M_{N,N+1}$, and also correspond with equations (2-40a) and (2-40b).

A solution for the $2(N + 1)$ unknown amplitudes E_v^i and E_v^r can be found from the $2(N + 1)$ equations included in the matrix expression (2-46). First, the two amplitudes E_{N+1}^i for the substrate and E_0^r for the vacuum have to be determined after connecting them by a matrix multiplication:

$$\begin{bmatrix} 1 \\ E_0^r \end{bmatrix} = \prod_{v=0}^{N} M_{v,v+1} \begin{bmatrix} E_{N+1}^i \\ 0 \end{bmatrix} \qquad (2\text{-}48)$$

With the known amplitude E_{N+1}^i for the substrate, the remaining amplitudes for all layers can then be calculated in accord with expression (2-46) in steps backward. Beginning with the known amplitude E_0^r for the vacuum, the remaining amplitudes can be calculated also in steps forward.

The primary intensity for a given glancing angle α_0 in the depth z of layer v is then given by

$$I_{\text{int},v}(\alpha_0,z) = I_0 \cdot |E_v^i \exp(-ik_0\alpha_v z) + E_v^r \exp(ik_0\alpha_v z)|^2 \qquad (2\text{-}49)$$

for z-values between $-d_v/2$ and $+d_v/2$.

By means of the primary intensity $I_{\text{int},v}$ the fluorescence intensity of an element x in the layer v can now be determined. The element x may be present in the vth layer with a mass fraction $c_{x,v}$. Its atoms may be excited to fluorescence in the primary wave-field $I_{\text{int},v}$ by photons of the energy E_0, i.e., by monochromatic radiation. At these conditions, the fluorescence intensity of the layer can be derived from equation (2-43) by replacing the fundamental parameters of the first layer by those of the vth layer. This original fluorescence radiation with a photon energy $E < E_0$ will further be absorbed on its way to the detector, which again should be mounted perpendicular to and close to the stratified medium. Absorption takes place in all layers above the relevant layer v, i.e., in the $(v-1)$ layers, with an index $j = 1, \ldots, v-1$. Consequently, the detector will measure an intensity that results from equation (2-43) by multiplication with an absorption factor

$$A_v = \exp\left[-\sum_{j=1}^{v-1} (\mu/\rho)_{j,E}\rho_j d_j \right] \qquad (2\text{-}50)$$

where $(\mu/\rho)_{j,E}$ is the mass-attenuation coefficient of the jth layer for photons of the energy E.

The fluorescence intensity to be measured by the detector can then be calculated by

$$\begin{aligned}
I_{x,v}(\alpha_0) = I_0 c_{x,v} S_{x,E_0} \varepsilon_{\det} T_{\text{air}} \, A_v \cdot &\left\{ |E_v^i|^2 \cdot \frac{1 - \exp(-[(\mu/\rho)_{v,E_0}/\alpha_v + (\mu/\rho)_{v,E}]\rho_v d_v)}{[(\mu/\rho)_{v,E_0}/\alpha_v + (\mu/\rho)_{v,E}]} \right. \\
&+ |E_v^r|^2 \cdot \frac{1 - \exp(-[-(\mu/\rho)_{v,E_0}/\alpha_v + (\mu/\rho)_{v,E}]\rho_v d_v)}{[-(\mu/\rho)_{v,E_0}/\alpha_v + (\mu/\rho)_{v,E}]} \\
&+ \left. 2Re\left[E_v^i \cdot E_v^r \cdot \frac{1 - \exp(-[-2ik_0\alpha_v/\rho_v + (\mu/\rho)_{v,E}]\rho_v d_v)}{[-2ik_0\alpha_v/\rho_v + (\mu/\rho)_{v,E}]} \right] \right\} \qquad (2\text{-}51)
\end{aligned}$$

This rather complex expression may even be summed up for an element x which is present in different layers v. It is evident that a polychromatic instead of a monochromatic excitation would further complicate this expression. It may be emphazised that the equation cannot be solved directly for any of the possibly unknown parameters $c_{x,v}$, ρ_v, or d_v. The mass fraction $c_{x,v}$ is even concealed in the mass-attenuation coefficients $(\mu/\rho)_{v, E_0}$ and $(\mu/\rho)_{v, E}$ and not only present as a single factor.

One special and simplified case should be mentioned, namely, that of a periodic multilayer. For this, only two different layers and consequently only two different transfer matrices are relevant. Any period of a double layer is characterized by one and the same characteristic matrix. On account of the periodicity, the total multilayer with N periods is characterized by the Nth power of this matrix. All further calculations are considerably shortened under this condition.

REFERENCES

1. Born, M., and Wolf, E (1980). *Principles of Optics*. Pergamon, Oxford; 6th ed., 1993.
2. Röseler, A. (1990). *Infrared Spectroscopic Ellipsometry*. Akademie-Verlag, Berlin.
3. de Boer, D.K.G. (1991). *Phys. Rev. B* **44**, 498.
4. de Boer, D.K.G., and van den Hoogenhof, W.W. (1991). *Adv. X-Ray Anal.* **34**, 35.
5. Kiessig, H. (1931). *Ann. Phys. (Leipzig)* **10**, 769.
6. de Boer, D.K.G., Leenaers, A.J.G., and van den Hoogenhof, W.W. (1995). *X-Ray Spectrom.* **24**, 91.
7. Huang, T.C., and Parrish, W. (1992). *Adv. X-Ray Anal.* **35**, 137.
8. Bedzyk, M.J., Bommarito, G.M., and Schildkraut, J.S. (1989). *Phys. Rev. Lett.* **62**, 1376.
9. Zheludeva, S.I., Kovalchuk, M.V., Novikova, N.N., and Sosphenov, A.N. (1995). *Adv. X-Ray Chem. Anal. Jpn.* **26s**, 181.
10. Barbee, T.W., Jr., and Warburton, W.K. (1984). *Mater. Lett.* **3**, 17.
11. Batterman, B.W. (1963). *Phys. Rev.* **133**, A759.
12. Batterman, B.W., and Cole, H. (1964). *Rev. Mod. Phys.* **36**, 681.
13. Schwenke, H., Berneike, W., Knoth, J., and Weisbrod, U. (1989). *Adv. X-Ray Anal.* **32**, 105.
14. Schwenke, H., and Knoth, J. (1993). "Total Reflection XRF." *In Handbook on X-Ray Spectrometry*. R. van Grieken and A. Markowicz (eds.), Practical Spectroscopy Series, Vol. 14, p. 453. Dekker, New York.
15. Klockenkämper, R., and von Bohlen, A. (1989). *Spectrochim. Acta* **44B**, 461.
16. Schwenke, H., Knoth, J., and Weisbrod, U. (1991). *X-Ray Spectrom.* **20**, 277.
17. de Boer, D.K.G. (1991). *Spectrochim. Acta* **46B**, 1433.
18. Gutschke, R. (1991). Diploma thesis, University of Hamburg.
19. de Boer, D.K.G., and van den Hoogenhof, W.W. (1991). *Spectrochim. Acta* **46B**, 1323.

20. Parratt, L.G. (1954). *Phys. Rev.* **95**, 359.

21. Abelès, F. (1950). *Ann. Phys. (Paris)* [12] **5**, 596, 706.

22. Król, A., Sher, C.J., and Kao, Y.H. (1988). *Phys. Rev. B* **38**, 8579.

23. Iida, A. (1992). *Adv. X-Ray Anal.* **35**, 795.

24. Sakurai, K., and Iida, A. (1992). *Adv. X-Ray Anal.* **35**, 813.

25. Kregsamer, P. (1991). *Spectrochim. Acta* **46B**, 1333.

26. Hüppauf, M. (1993). Doctoral thesis, RWTH Aachen, and JÜL-report JÜL-2730, ISSN 0366–0885.

27. Holz, Th. (1992). Diploma thesis, Technical University, Dresden.

28. Weisbrod, U., Gutschke, R., Knoth, J., and Schwenke, H. (1991). *Appl. Phys. A* **53**, 449.

29. Weisbrod, U., Gutschke, R., Knoth, J., and Schwenke, H. (1991). *Fresenius' J. Anal. Chem.* **341**, 83.

30. Schwenke, H., Gutschke, R., and Knoth, J. (1992). *Adv. X-Ray Anal.* **35**, 941.

31. Lengeler, B. (1992). *Adv. X-Ray Anal.* **35**, 127.

32. Schwenke, H., Gutschke, R., Knoth, J., and Kock, M. (1992). *Appl. Phys. A* **54**, 460.

33. de Boer, D.K.G., and Leenaers, A.J.G. (1995). *Adv. X-Ray Chem. Anal. Jpn.* **26s**, 119.

34. Kawamura, T., and Takenaka, H. (1994). *J. Appl. Phys.* **75**, 3806.

35. Veigele, W.J. (1973). *At. Data Tables* **5**.

36. Bertin, E.P. (1975). *Principles and Practice of X-Ray Spectrometric Analysis.* Plenum Press, New York.

37. Hubbel, J.H., Veigele, W.J., Briggs, E.A., Brown, R.T., Cromer, D.T., and Howerton, R.J. (1975). *J. Phys. Chem. Ref. Data* **4**, 471.

INSTRUMENTATION FOR TXRF

Most X-ray fluorescence spectrometers today operate in the wavelength-dispersive mode, but the number of energy-dispersive spectrometers is rapidly increasing. Wavelength-dispersive instruments are estimated at about 15,000 worldwide, whereas there are some 3000 units of the energy-dispersive type. Total-reflection devices are only efficient and available in the energy-dispersive version, and there may be about 300 presently in operation.

The first commercially available TXRF instrument was built in 1980 (EXTRA, by Rich. Seifert & Co., Ahrensburg, Germany). It was protected by a patent of Marten, Rosomm, and Schwenke [1]. The successive improved model, EXTRA II a, is distributed today by Atomika Instruments (Oberschleissheim, Germany). This company also sells model TXRF 8010, especially suitable for the examination of wafers. Since 1988, two Japanese companies have put instruments on the market: the TREX series is built by Technos, Osaka (represented in Europe by Philips, Eindhoven, the Netherlands); the 3700 series is constructed by Rigaku, Osaka (represented in Europe by Rigaku Europe GmbH, Düsseldorf, Germany). These instruments are compact, self-contained units consisting of a power supply, an X-ray tube, a special filter or even a monochromator, a sample chamber, a detector, and a multichannel analyzer. The optical path of X-rays is determined by apertures and is safeguarded by lead shielding. These compact instruments are licensed only as highly protective units that switch off the X-ray tube if the protective shielding is removed.

Modular equipment of great versatility has also been constructed consisting of a collimator system, a special filter or monochromator, a sample-holder device, and connection flanges for an X-ray tube and a detector. This module is distributed by the IAEA (International Atomic Energy Agency, represented by P. Wobrauschek, Vienna). The generator, X-ray tube, detector, and multichannel analyzer have to be supplied elsewhere and coupled to this module.

Maintenance and operating costs of a TXRF device are relatively low. The generator needs about 3 kW electric power, the X-ray tube has to be cooled by tap water (5 L/min), and the detector must be cooled by liquid nitrogen (10 L/week). If an in-door closed water-cooling cycle is available, water consumption can be greatly limited. In principle, no vacuum is necessary and no gases are needed. The consumption of sample carriers is small. Quartz glass carriers, for example, can easily be cleaned after application and reused. Plexiglas carriers are disposable, each costing only pennies.

In general, all compact and self-contained units are easy to operate. The menu-driven setting of the power supply, as well as automatic changing of samples and recording of the spectra, all guarantee a user-friendly mode of operation. If two X-ray tubes are provided, a simple switchover of the power supply avoids the need for troublesome tube changing. Sophisticated software is available to facilitate calibration, evaluation, and storage of data.

3.1. BASIC INSTRUMENTAL SETUP

In Section 2.2, it was pointed out in detail that standing waves appear when X-rays interfere at external total reflection. The standing waves may arise in front of a thick and flat substrate and/or within a layered structure on top of such a substrate. Sample material placed in the field of standing waves can be excited to X-ray fluorescence. Two cases may be distinguished in general: (1) For a granular residue, the fluorescence intensity will be constant, i.e., independent of the glancing angle if this angle is reduced below the critical angle of total reflection. The spectral background arising from Rayleigh and Compton scattering of the primary beam is likewise constant. (2) For thin layers, however, the fluorescence intensity will be angle dependent—the layers may be buried in or deposited on a thick substrate and may be self-reflecting or not.

In order to take TXRF measurements, great demands have to be made on the instrumentation:

- The glancing angle of incidence for the primary beam must be rather small in order to ensure external total reflection. The critical angle is on the order of 0.1° for primary X-rays of some 10 keV as normally applied (see Table 1-3).
- The primary beam should be shaped like a strip of paper, realized by an X-ray tube with a line focus. Apertures should restrict the beam to some 10 μm in height and about 10 mm in width since the detector window is usually smaller than 1 cm in diameter.

For the examination of a granular residue, case 1, only a single measurement at a fixed angle setting has to be carried out. Two conditions have to be met:

- The glancing angle must be set to about 70% of the critical angle of total reflection. Depending on excitation energy and carrier material, it should be fixed to maybe 0.07°. The divergence of the primary beam should be restricted to only 0.01°. For a stable setting of such small angles, the instrumental arrangement has to be very solid and compact.

- Intensive spectral peaks or a broad band of the primary brems-continuum should first be selected for excitation. The high-energy part of the primary spectrum must be eliminated by a filter so that total reflection can occur at a small but not too small angle according to equation (1-34). This high-energy part of the spectrum would not be totally reflected under the larger glancing angle but would lead to an increased background. A prior low-pass or bandpass filtration prevents this detrimental effect.

For the examination of thin layers, case 2, not only a single measurement at a fixed angle is needed but an angle-dependent intensity profile has to be recorded. Instead of the aforementioned conditions, the following requirements have to be met:

- The angle of incidence must be capable of being varied between 0° and 2° in steps of 0.01°. This can be realized by tilting the sample around an axis located on the surface. Great accuracy of the fine-angle control is necessary for quantification and should be better than 0.005° absolutely.
- A strong spectral peak of the primary spectrum should be selected by a suitable monochromator. Only a monochromatic incident beam produces the angle-dependent intensity profiles that are distinctly determined by the layered samples. Additional spectral parts, especially strong peaks, would blur the distinct correlation.

The basic design of TXRF instrument based on these conditions is demonstrated in Figure 3-1. The primary beam is generated by a high-power X-ray tube with a line focus. It may either be a fixed or a rotating anode tube. In order to increase certain peaks in relation to the spectral continuum, thin metal foils are easily placed in front of the X-ray tube. They work as filters, as mentioned in Section 1.3.3. By means of a pair of precisely aligned diaphragms or slits, the beam will be shaped like a strip of paper.

Since the mid-1980s, a synchrotron beam has sometimes been used for excitation [2–4]. This X-ray source certainly provides an ideal primary beam with natural vertical collimation, polarization, and high brilliance. However, it requires a large-scale piece of equipment, the synchrotron, which is not widely available for research work and even less so for routine analyses (see Section 6.2.1.2).

The polychromatic beam of conventional X-ray tubes is deflected by what is called the *first reflector*, and which alters the primary spectrum. For trace analyses of granular residues, a simple quartz glass block is sufficient in that role, acting as a totally reflecting mirror or low-pass filter. It only cuts off the high-energy part of the brems-continuum under grazing incidence. For surface

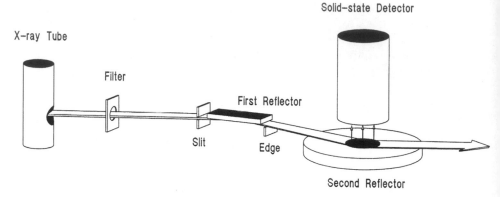

Figure 3-1. Basic design of a TXRF instrument. This diagram differs from Figure 1-1b mainly by the addition of the first reflector.

and thin-layer analyses, the first reflector has to be a real monochromator. Generally, natural crystals or multilayers are used, acting as Bragg reflectors (as described in Section 2.1.2).

After passing this first reflector, the primary beam hits the sample carrier as the *second reflector* under grazing incidence. The sample carrier may be loaded with some sample material or may be exchanged by a layered or unlayered substrate representing the actual object of analysis itself. For trace analyses, the carrier may simply be pressed against two parallel cutting edges or three steel balls so that a fixed angle setting is guaranteed. For surface- or thin-layer analyses, however, a special device is needed for the positioning and tilting of the sample. This component has to control the angle of incidence, which is the governing free variable for angular intensity profiles.

The fluorescence intensity of the sample is generally recorded by an energy-dispersive solid-state detector, usually a Si(Li) detector. It is mounted perpendicular to the carrier plane to obtain spectra with a minimum scattered background (as discussed in Section 1.3.2). The distance to the sample is reduced to about 1 mm in order to secure the detection of the fluorescence radiation within a large solid angle. This intensity is registered by a multichannel analyzer, leading to an energy-dispersive spectrum. Measurements are usually carried out in ambient air. In order to suppress the Ar-peak from ambient air, the measuring chamber can be flushed with nitrogen. For the detection of low-energy peaks of low-Z elements, a helium flush or even an entire vacuum chamber are recommended (see Section 6.2.1.1).

3.2. THE HIGH-POWER X-RAY SOURCE

A strip-like X-ray beam is needed for excitation, showing a high intensity in a certain spectral region. Consequently, a high-power X-ray source must be chosen for TXRF, just as it is used for X-ray diffraction (XRD). But the emitted X-ray beam must still be adapted with respect to geometry and chromaticity.

The applied X-ray sources consist of a generator and a fine-focus X-ray tube, also known as fine-structure tube. The generator supplies the X-ray tube with high-potential power for the anode and filament power for the cathode. Normally, the very stable generators of conventional XRF or XRD are applied. They deliver a rectified high-voltage between 5 and 60 kV (or even 100 kV) and a direct current of 5 to 80 mA (or even 100 mA), usually in increments of 1 kV and 1 mA, respectively. The output power can reach a maximum load of 3 or 4 kW. High-quality generators are stable to $\pm 0.01\%$ for both voltage and current provided that the line fluctuations are less than $\pm 10\%$. The usual electrical safety precautions are generally taken.

A new generation of high-powered units can provide a maximum load of 18 kW (or even 30 kW), a maximum voltage of 60 kV, and a maximum current of 300 mA (or even 500 mA). The stability of these generators is $\pm 0.1\%$, which is indeed remarkable, but they need to be cooled by a water flow rate of about 6 L/min.

3.2.1. Fine-Focus X-ray Tubes

X-ray tubes provide the primary X-ray beam by which the sample is excited to fluorescence. Fine-focus X-ray tubes with a fixed anode are generally used for TXRF [5]. As shown in Figure 3-2, they consist of a spiral filament acting as the cathode and a water-cooled block of copper as the anode. Both electrodes are sealed off in an evacuated glass-metal cylinder. In contrast to Figure 1-2, the anode is not a beveled block but a right cylinder with a horizontal plane. The filament made of tungsten wire is embedded in a narrow steel groove, 1 mm wide and about 10 mm long. The copper block is plated with the actual anode material such as chromium or tungsten. The cathode is operated at high negative potential, while the anode is earthed.

When the filament is fed by the heating current at white heat, it emits electrons. They are attracted and accelerated in the direction of the anode at a distance of about 20 mm. The bombarded area of the target is about 0.25 mm × 10 mm, representing a line focus of X-rays. These X-rays are emitted in all directions but only emerge through a thin side window. If the diameter is about 6 mm and the window-to-spot distance is nearly 30 mm, a beam will leave the tube as a cone with an aperture of about 12°. If the total

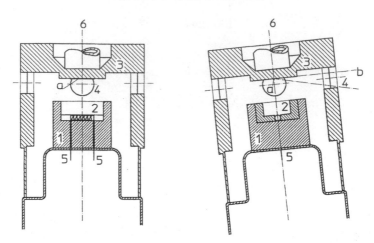

Figure 3-2. Fine-focus X-ray tube in two sectional views: (1) cathode block; (2) tungsten filament; (3) anode block; (4) thin window; (5) electrical connections; (6) cooling-water connection; (a) line focus; (b) radiation cone. (Schematic representation of tube SF 60 produced by Rich. Seifert & Co., Ahrensburg, Germany.)

tube is now tilted by $6°$, the beam axis can run horizontally. On the outside, the focus is *observed* under the small angle of $6°$ and *appears* as a line of $25\,\mu m \times 10\,mm$.

Fine-focus X-ray tubes are available with several different target materials: Au, W, Ag, Mo, Cu, Co, Fe, and Cr (listed with decreasing atomic number). The maximum permissible power is about $2\,kW$ for high-Z anodes and about $1\,kW$ for the low-Z ones. The exit window is made of beryllium, which is highly transparent for X-rays. Foils of $0.2–1$ mm thickness are used, while the high-Z anode tubes require the thicker windows. The water consumption is about $4\,L/min$ at a pressure of $3–5$ bar and a temperature of $20–30\,°C$. The tubes' life span is some $3000–6000$ operational hours.

In order to achieve such a long life, some rules should be observed. Loading of the tube as well as switching it off should never be performed suddenly but only as a careful step-by-step operation. The warm-up should take between 15 and 45 min, depending on the period of interruption, and the cooling-down phase should last for 10 min. The maximum power should never be exceeded. Voltage and current should be chosen so that their product is always less than a maximum rating. All this should be performed under automatic control. Tube changes, however, have to be carried out manually with utmost care.

Modern X-ray tubes are perfectly shielded and permit operation without radiation hazard. Nevertheless, participation in a monitoring program is recommended.

3.2.2. Rotating Anode Tubes

Conventional X-ray tubes have a fixed anode and are sealed off under vacuum. In contrast to these tubes, a new class of tubes is manufactured with a rotating anode, as shown in Figure 3-3. The anode is bombarded by electrons on its rotating sidewall surface. The rotary shaft is sealed by a special gasket, and a high vacuum ($< 10^{-4}$ Pa) is maintained by a pump system consisting of a turbo-molecular pump with an oil-rotary backing pump. Rotating anode tubes can be operated with a maximum power of 18 and even 30 kW at a voltage of 60 kV and a current of 300 and even 500 mA [6]. Consequently, they provide an X-ray source of high brilliance. The emergent primary beam has a 9-fold or even 15-fold intensity compared to a conventional primary beam of a 2 kW tube.

The high thermal stress of the anode is reduced by rotation at 2500–6000 rpm and by intensive water cooling with a flow rate of 8–15 L/min at a pressure of 2–3 bar. The size of the line focus typically amounts to 0.2×10 mm^2, just as for conventional tubes. The stability is about 0.1% for voltage and current, not very much poorer than for conventional tubes. Of course, this high-performance X-ray source with a high-power generator, a rotating anode device, and a vacuum system is much more expensive than a conventional X-ray source. On the other hand, the new tubes are highly flexible because the anodes can be interchanged. Several anodes are available with target materials also applied for the former tubes. These exchangeable components are of course much cheaper than an entire conventional tube.

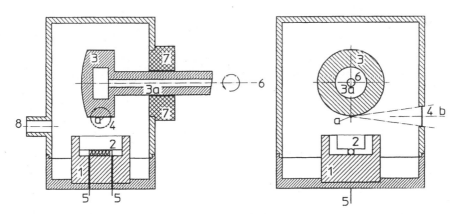

Figure 3-3. Rotating anode tube in two sectional views: (1) cathode unit; (2) tungsten filament; (3) cylindrical anode with (3a) rotary shaft; (4) thin window; (5) electrical connections; (6) cooling-water connection; (7) sealing gasket; (8) high-vacuum flange; (a) line focus; (b) radiation cone. (Sketched after a diagram in a technical brochure of Rigaku International Corporation [6].)

Rotating anode tubes are warranted for 2000 h of service-free operation. A safety circuit provides for automatic shutdown in case of overloads and possible malfunctions. The rules given for running the conventional X-ray units should be observed nonetheless.

3.3. THE BEAM-ADAPTING UNIT

The primary X-ray beam must still be adapted with respect to its geometric shape and spectral distribution. Shaping the beam is easily done by two collimator slits or at least by two metallic edges acting as diaphragms. Silver, steel, and platinum are generally the metals chosen, with a thickness of about 1 mm and a width of about 20 mm. They are either fixed in entirely self-contained devices or have to be adjusted in modular components. In any case, their task is to mask out a strip-like beam of some 10 μm thickness.

Alteration of the spectral distribution is not as easy to do as is shaping. A metal foil is usually employed as a supplementary means of accomplishing this, but the real goal is achieved only by a first reflector. For *trace analysis*, a quartz glass mirror is preferably used, acting as a low-pass filter [7]. But a multilayer can also be used, acting as a monochromator or rather as a broad bandpass filter. Even two components or combinations of them may be utilized. The combination quartz–multilayer and multilayer–multilayer was shown to be highly suitable [8], whereas a crystal monochromator is obviously unsuited [9]. The single or double multilayer arrangement allows selection of a specific excitation energy without the need to change the X-ray tube. This energy tuning can be performed by a vertical shifting and/or a horizontal displacing of the multilayers.

For *surface and thin-layer analyses*, a natural crystal acting as a true monochromator or small bandpass filter may be applied as a first reflector. But multilayers have also been used successfully (by Atomika and Rigaku). They may possibly be chosen as a combination of two.

3.3.1. Low-Pass Filters

In order to achieve total reflection at the sample carrier, the primary beam must strike this carrier as a second reflector at a small angle of incidence. For X-ray photons of 60 keV, the glancing angle should be set at only 0.015°, while the divergence of the beam must be held to an even smaller angle. Such a narrow primary beam would provide only poor intensity. Besides, the high-energy photons would scarcely be effective since excitation is preferably induced by lower energy photons, e.g., by the characteristic radiation of the anode material. For this reason, the glancing angle and the divergence of the primary beam are increased threefold. In that case, the high-energy photons would not be totally reflected but would partly be scattered, causing a significant background. To avoid this effect, the high-energy photons must be eliminated first. This can be done by a first reflector acting as a low-pass or cutoff filter. It only transmits the low-energy part of the primary spectrum but eliminates the high-energy part [7, 10].

Such a filter can be realized by a simple quartz glass block. It represents an ideal low-pass filter if applied as a totally reflecting mirror. At a given angle of incidence, low-energy photons are totally reflected whereas high-energy photons are absorbed or scattered. This angle is called the cutoff angle. If a cutoff is wanted for a certain energy E_{cut}, a corresponding cutoff angle α_{cut} has to be set. In accordance with equation (1-34), it is determined by

$$\alpha_{cut} = \frac{1.65}{E_{cut}} \sqrt{\frac{Z}{A}\rho} \qquad\qquad (3\text{-}1)$$

where E_{cut} is given in keV and ρ in g/cm^3 in order to get α_{cut} in degrees. For quartz glass, this relation can be specified by

$$\alpha_{cut} = \frac{1.73}{E_{cut}} \qquad\qquad (3\text{-}2)$$

Obviously, other materials can also be used as totally reflecting mirrors; however, quartz glass is highly effective and easy to machine. The efficiency of quartz glass mirrors is demonstrated by Figure 3-4. The reflectivity as a function of the photon energy is shown for three different cutoffs. The steplike decrease from nearly 100% to about 0% with only weak tailings is evident. The low-energy part is preserved, whereas the high-energy part is annihilated to a large extent. Deviations from the ideal cutoff are caused by absorption.

In practice, an assembly of filters and diaphragms can be arrayed in a different manner. Figure 3-5 shows a simple arrangement with a single reflector or mirror, one front slit, and one back cutting edge. In contrast to this design, the devices of Figures 3-6 and 3-7 both employ a double reflector, one front slit, and one back edge or a second slit. They mainly differ from one another by the reflector length allowing either a twofold reflection (Figure 3-6) or a threefold reflection (Figure 3-7). The reflection blocks are made of quartz glass (mostly Suprasil) and are about 10 mm thick, 15 mm wide, and 30–50 mm long in the version of Figure 3-6. They are mounted in a three-point bearing with three adjustable steel balls and are removable from their pedestal. The double reflector of Figure 3-7 is 100 mm long and is joined together with two spacers between the reflector blocks [11, 12]. It can only be adjusted as

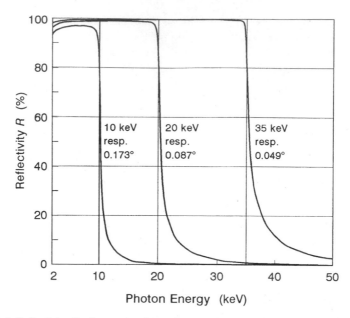

Figure 3-4. Reflectivity R of a quartz glass mirror acting as a low-pass filter. Three different settings of the cutoff energy or the cutoff angle are represented.

a whole. Instead of a plane reflector, a slightly curved mirror can be applied for the simple arrangement of Figure 3-5. With a 50 m radius of curvature, the signal intensity is thus increased fivefold due to a larger divergence of the primary beam [13].

In all three cases, the primary spectrum can first be altered by a simple metal foil. Different metals with a thickness of some 10–200 μm can each be mounted on a frame and positioned in front of the X-ray tube. They can easily be exchanged without any special adjustment. As noted in Section 1.3.3, certain spectral regions are attenuated thereby as a supplement to the low-pass effect of the first reflector.

All three arrangements may be prealigned as regards the reflector and collimator system, whereas the X-ray tube should be adjusted last. For this purpose, the tube is first shifted vertically and then tilted horizontally until a maximum intensity of the fluorescence signal is reached. This procedure is only necessary after installation of the total equipment or exchange of individual components.

An evaluation of the three systems has to consider the divergence, the intensity, and the spectral distribution of the adapted primary beam. Obvious-

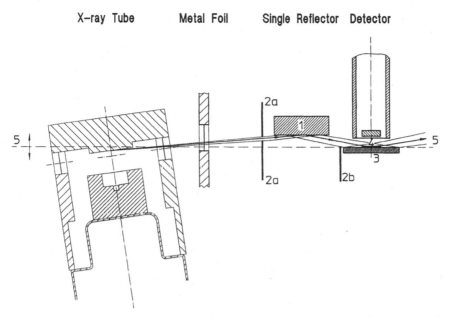

Figure 3-5. A single reflector acting as a totally reflecting mirror and a low-pass filter as well. Only the low-energy part of the primary tube spectrum passes through this filter and is used for excitation in TXRF: (1) single reflector; (2a) front slit; (2b) back diaphragm; (3) sample carrier; (4) sample; (5) reference plane. (Schematic representation of EXTRA II produced by Rich. Seifert & Co., Ahrensburg, Germany.)

ly a smaller divergence is realized with the double reflector, especially at threefold reflection. Consequently, the spectral cutoff is sharper, on the one hand, but the total intensity is reduced, on the other. Priority should be given to the higher intensity obtained with the single reflector. Besides, this simple version is easier to construct and to adjust as well.

The simple arrangement of Figure 3-5 will now be described in further detail. We define the horizontal distance of the X-ray tube to the first reflector as a_1 and the distance of the first reflector to the second reflector or the sample carrier as a_2, in both cases with reference to the respective centers. The carrier will be hit at the defined angle α_2 if the first reflector is lifted above the reference plane by

$$h_1 = a_2 \tan \alpha_2 \tag{3-3}$$

Furthermore, the first reflector must be tilted by an angle $\alpha_1 - \alpha_2$ with respect to the reference plane so that the X-ray beam will be reflected under the chosen

X−ray Tube   Metal Foil   Double Reflector   Detector

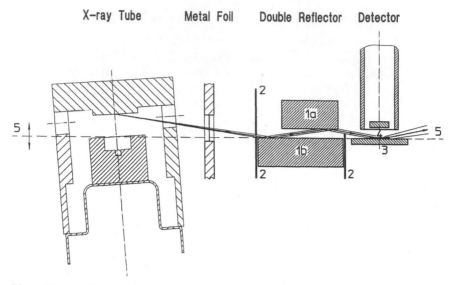

Figure 3-6. Double reflector acting as a low-pass filter used for TXRF. The beam-limiting device enables a twofold reflection at the upper and lower blocks of the reflector before the primary beam reaches the sample: (1a) upper reflector block; (1b) lower reflector block; (2) front slit and back diaphragm; (3) sample carrier; (4) sample; (5) reference plane. (Schematic representation of an early EXTRA II produced by Rich. Seifert & Co., Ahrensburg, Germany.)

angle α_1. The line focus of the X-ray tube must be lowered by

$$h_0 = a_1 \tan(2\alpha_1 - \alpha_2) - h_1 \qquad (3\text{-}4)$$

In addition, the X-ray tube must be tilted by $2\alpha_1 - \alpha_2 + 6°$.

The carrier will be irradiated with a width w_2 if the X-ray beam is limited to a thickness or height t at the sample position. This thickness must be

$$t = w_2 \sin \alpha_2 \qquad (3\text{-}5)$$

The width w_2 should be chosen smaller than the diameter of the sample carriers but wider than that of the detector crystal.

The thickness t of the X-ray beam determines the aperture or divergence of this beam. It can be calculated by

$$\Delta\alpha = \frac{w_2 \sin \alpha_2}{a_1/\cos(2\alpha_1 - \alpha_2) + a_2 \cos \alpha_2} \frac{180°}{\pi} \qquad (3\text{-}6)$$

X-ray Tube   Metal Foil   Double Reflector   Detector

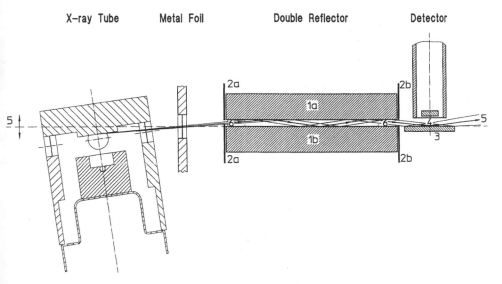

Figure 3-7. Double reflector acting as a low-pass filter used for TXRF. The beam-adapting device allows a threefold reflection of the primary beam before excitation of the sample: (1a) upper reflector block; (1b) lower reflector block; (2a) front slit; (2b) back slit; (3) sample carrier; (4) sample; (5) reference plane; (6) spacer. (After Schwenke and Knoth [10].)

For small angles α_1 and α_2, it can be approximated by

$$\Delta\alpha \approx \frac{w_2}{a_1 + a_2}\alpha_2 \tag{3-7}$$

This quantity is independent of α_1 and only dependent on α_2. Finally, the irradiated area of the first reflector can be determined by the width

$$w_1 = w_2\frac{\sin\alpha_2}{\sin\alpha_1} \tag{3-8}$$

For example, a Mo-tube may be chosen as the X-ray tube and quartz glass as the reflector material: α_2 is set to 0.07° in order to ensure total reflection at the sample carrier (recommended below in Section 4.4.2); α_1 is set to 0.09° in order to cut off the spectrum above 20 keV by means of the first reflector. The signal intensity is optimized by maintenance of close distances between the X-ray tube, the first reflector, and the sample carrier. If the distances a_1 and a_2 are kept to a minimum of 100 and 40 mm, respectively, the first reflector has to be raised by $h_1 = 49\ \mu m$ and the X-ray focus lowered by about $h_0 = 143\ \mu m$. The

X-ray beam must be limited to $t = 24\,\mu$m in height, determining an aperture of $\Delta\alpha = 0.01°$. In this case, the first reflector is irradiated within a width of $w_1 = 16\,$mm while the second reflector, i.e., the sample carrier, is irradiated within a width $w_2 = 20\,$mm.

3.3.2. Monochromators

For surface and thin-layer analyses, intensity profiles have to be recorded that are uniquely dependent on the glancing angle and not on the photon energy. Consequently, considerable demands on the spectral purity have to be made. The primary beam needs to be fairly monochromatic, which can be realized neither by a low-pass filter nor by a foil filter but only by a monochromator. But even trace analyses may profit from monochromatic excitation, as already mentioned. For such cases, the first reflector of Figure 3-5 must be a natural crystal or a multilayer. Both types of monochromators are used as Bragg reflectors, with a definite energy band selected at a particular angle of reflection, as discussed in Section 2.1.2.

This angle should be set in accord with equation (2-8). For a chosen photon energy E in keV, the angle can be calculated by

$$\alpha = \arcsin\left(\frac{0.620}{Ed}\right) \tag{3-9}$$

where d is the interplanar spacing (in nm) of the reflector in use. A lithium fluoride, LiF(200), or a graphite crystal, C(002), can be used; the latter should preferably be a highly oriented pyrolytic graphite (HOPG). But a set of multilayers is also available, e.g., W/Si, W/C, and Mo/B$_4$C (manufacturer: Ovonic Synthetic Materials Co. Inc., Troy, Michigan). They are used to select a strong K- or L-peak of the anode material of an X-ray tube, e.g., the Mo-$K\alpha$ or W-$L\beta$ peak. Corresponding d-values and glancing angles are listed in Table 3-1.

The Bragg angle for common multilayers is about $1°$ and for natural crystals is on the order of $10°$ due to the smaller spacing d. As a consequence, the X-ray tube of Figure 3-5 has to be lowered by about 3.3 mm ($\alpha_1 = 1°$) or even 3.6 cm ($\alpha_1 = 10°$), in accord with equation (3-4). The irradiated area of the first reflector thus shrinks to about 2.0 mm ($\alpha_1 = 1°$) or 0.2 mm ($\alpha_1 = 10°$), in accord with equation (3-8). Apart from the lowering of the tube and the larger tilt of both the tube and the first reflector, the geometric arrangement of Figure 3-5 can be maintained in principle. The variation of the glancing angle α_2 will be treated in Section 3.4.3.

Of course, the various crystals and multilayers differ in efficiency when acting as a monochromator [14]. The decisive characteristics are the peak

Table 3-1. Natural Crystals and Multilayers Frequently Used to Select the Mo-$K\alpha$ or W-$L\beta$ Line for Excitation in TXRF

Crystal	LiF(200)	C(002)	W/Si	Mo/B$_4$C
Spacing d (nm)	0.2014	0.3354	2.5	8.0
Bragg angle α:				
For Mo-$K\alpha$	10.15°	6.08°	0.814°	0.254°
For W-$L\beta$	18.55°	11.01°	1.469°	0.459°
Peak:				
Reflectivity	40–50%	ca. 20%	80–90%	ca. 80%
Bandwidth	5–25 eV	40–200 eV	400–1200 eV	200–800 eV

reflectivity and the spectral bandwidth. Raw data for these two quantities are listed in Table 3-1. The peak reflectivity is some 10% for natural crystals, about 90% for multilayers, but nearly 100% for a totally reflecting mirror or low-pass filter. The bandwidth shows a similar trend. It can be estimated from

$$\delta E = E \frac{\delta\alpha}{\tan\alpha_1} \qquad (3\text{-}10)$$

where δE is the bandwidth; E, the selected photon energy; α_1, the adjusted angle of the first reflector; and $\delta\alpha$, the decisive divergence of the X-ray beam. This divergence is determined by the aperture of the beam ($\Delta\alpha$ about 0.01°) but also by a certain misorientation of the mosaic structure of natural crystals and by a certain variance of the spacing of multilayers. The bandwidth of natural crystals has been estimated as ranging from about 5 to 200 eV and that of multilayers from about 200 to 1000 eV.

Consequently, multilayers can separate the $K\alpha$ and $K\beta$ or the $L\alpha$ and $L\beta$ peaks of a primary beam but not the doublets $K\alpha_1, K\alpha_2$ or $L\alpha_1, L\alpha_2$. These doublets, however, can be separated and selected by most natural crystals excepting HOPG. In consequence, natural crystals will be used if a high spectral selectivity is needed in preference to intensity. The lower integral intensity might be compensated by a powerful X-ray source, e.g., a rotating anode tube. Multilayers provide a superior intensity at the expense of selectivity. Nevertheless, they can be applied for surface and thin-layer analyses if their selectivity is just sufficient for a particular $K\alpha$- or $L\alpha$-excitation.

Moreover, multilayers can be used as simple low-pass filters [12]. For that purpose, the glancing angle should be reduced below the critical angle of total reflection, i.e., from about 1° down to some 0.1°. The upmost layers will then act as a totally reflecting mirror with a low-pass effect. Multilayers thus have the advantage of being usable as both a monochromator and a low-pass filter.

3.4. SAMPLE POSITIONING

The sample to be analyzed can be presented in a small amount or volume on a solid and compact carrier. Various materials are in use as sample carriers. They have to be optically flat and even, in order to ensure the total reflection of X-rays. But they also should meet some further requirements, e.g., a high resistance to acids and solvents and a low price. There is no ideal carrier for *all* purposes but quartz glass and Plexiglas can be recommended for a lot of applications.

The loaded carriers are inserted into a sample changer, and from there they are sequentially brought into the fixed measuring position. When a sample changer is lacking, the samples are placed in this position manually and directly. Analyses are carried out at a fixed angle smaller than the critical angle of total reflection.

The sample to be investigated can also be presented as a flat disk that totally reflects the X-ray beam itself. Wafer material is mostly analyzed in this way. Generally, the wafers are loaded from a sample changer into the start position. Subsequently, they are slightly tilted for stepwise angle variation and recording of the fluorescence intensity.

3.4.1. Sample Carriers

For trace analyses of granular residues, a carrier is required that serves as a sample support and as a totally reflecting mirror. Of course, only solid-state media are suitable. They should be highly reflective because the spectral background is reduced in proportion to $1 - R$, in accord with equation (2-20). Besides, they have to be optically flat to ensure the high reflectivity, as predicted by theory. The roughness should be < 5 nm within an area of about 1 mm^2, corresponding to about $\lambda/100$, where λ is the mean wavelength of the visible light. The waviness should be $< 0.001°$ within an area of about 1 cm^2, corresponding to a radius of curvature of about 600 m. Commercially available carriers guarantee these characteristics. They can be checked by means of a profilometer and a contour meter.

Furthermore, carriers should be free of impurities so that no blank values appear for elements to be detected. The carrier material itself must not have a fluorescence peak in the spectral region to be considered. In addition, the carriers have to be chemically inert when strong inorganic acids and organic solvents are to be analyzed. The carriers should be easy to clean so that they

Table 3-2. Important Properties of Various Materials Used as Sample Carriers

Carrier Material	Plexiglas	Glassy Carbon	Boron Nitride	Quartz Glass
Critical angle for				
Mo-$K\alpha$	0.08°	0.08°	0.10°	0.10°
Reflectivity at 0.07°	99.8%	99.8%	99.9%	99.4%
Flatness	Good	Fair	Good	Excellent
Purity	Good	Fair	Good	Excellent
Fluorescence	None	None	None	Silicon
Resistance	Insufficient	Good	Excellent	Good
Cleaning	Not necessary	Difficult	Easy	Easy
Price for one[a]	$0.10	$30	$60	$28

Source: After Prange and Schwenke [19].

[a] U.S. currency.

can always be reused. Finally, they should be commercially available and inexpensive.

Several materials have been applied as sample carriers. Especially quartz glass (mostly Suprasil), polymethyl methacrylate (generally Perspex or Plexiglas) [15], glassy carbon [16], and boron nitride [17] have been used, but also one-element materials like silicon or germanium [18]. Table 3-2 gives an overview of different carrier materials [19].

Most carriers are on sale as circular disks with a diameter of 30 mm and a thickness of 2 or 3 mm, but rectangular carriers are also available. Their critical angle of total reflection amounts to about 0.1° for Mo-$K\alpha$ radiation, as shown in Table 3-2. The reflectivity is between 99.4 and 99.9% at a fixed glancing angle of 0.07°. According to equation (2-20), the spectral background should be *higher* by a factor of 3–6 for the *less* reflecting quartz glass in comparison to the other materials. However, this effect is compensated by a better flatness and purity of commercial quartz glass carriers. In addition, quartz glass is chemically resistant to most acids besides hydrofluoric acid.

The other materials still benefit from the absence of a Si-peak so that silicon and related light elements can be determined. The latter is not possible if quartz glass carriers are applied. The most resistant material is boron nitride, which is even suitable for the analysis of strong acids and solvents. Unlike these carriers, Plexiglas carriers are only applicable to aqueous solutions or suspensions. Plexiglas, however, is an extremely low-priced material, so that it is not necessary to clean and reuse these carriers. Glassy carbon is preferentially used for electrochemical applications because of its electrical conductivity [16]. In general, quartz glass and Plexiglas carriers are the carriers most used for micro- and trace analyses.

3.4.2.   Fixed-Angle Adjustment

The sample carriers are usually inserted into plastic holders and are either placed manually into their final measuring position or first loaded into a sample changer. From there they can be brought into the measuring position automatically one after another. The carriers are lightly pressed—preferably upward—either against two parallel cutting edges or four ball-points with a clearance of about 20 mm. These reference lines or points are fixed and define the plane of reflection. They also determine the angle of incidence for the primary beam. This beam should pass between the cutting edges or ball points and should be reflected under an angle of about 70% of the critical angle of total reflection. For quartz glass and Plexiglas carriers this angle amounts to about 0.07°. The fixed position of the carriers is the *only* position necessary for measurements, i.e., for recording the spectra.

A sample changer is recommended for the investigation of a lot or series of samples. Changers with a capacity of up to 35 carriers are commercially available. Sample changing can be carried out under computerized (PC) control. The whole device is usually incorporated in a plastic chamber for dust protection. This chamber can be flushed with a gas such as helium or nitrogen, and it can be evacuated if necessary.

3.4.3.   Stepwise-Angle Variation

For surface and thin-layer analyses, the sample has to be present as a flat disk capable for total reflection of the primary beam. Wafers are especially suited for this kind of investigation, e.g., Si-wafers or GaAs-wafers. For that purpose, sample holders are constructed that grip wafers with a diameter between 100 and 200 mm. The wafers may be stacked in a magazine and may be loaded into the sample holder one at a time and then set back after the measurement. This process can be carried out automatically, even by a robot.

The sample-positioning device is the core of those instruments that are suitable for surface and thin-layer analyses. The sample should be capable of being adjusted in the reference plane and of being tilted around a horizontal axis. The angle of incidence should be variable, preferably stepwise. This quantity is the key parameter for recording the angular dependent intensity profiles and is the basis for surface and thin-layer analysis. Apart from being adjustable, the sample should be displaceable in order to set each spot of a larger sample in the measuring position and to check the total surface.

Figure 3-8 schematically depicts a convenient device for angle variation and sample positioning with 6 degrees of freedom. But an even simpler device with ball-points and stepper motors may be sufficient. The wafer or sample S is placed on a flat carrier C (made of float glass). As an attachment, either

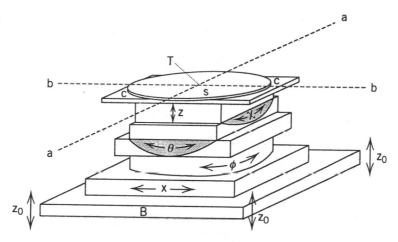

Figure 3-8. Schematic diagram of an angle variation and sample-positioning device needed for thin-layer analyses by TXRF: B = solid base; C = flat carrier; S = flat sample or wafer; a and b = horizontal axes spanning the reference plane; T = tilt center; x = lateral movement; ϕ = rotation on the vertical axis; Θ = tilt around axis a; χ = tilt correction for axis b; z = vertical shift of the sample; z_0 = vertical shift of the base.

a mechanical chuck is used, applying a slight pressure, or an electrostatic chuck is designed to apply a specific voltage. The sample can be adjusted to the tilt center T by a vertical shift z, and this center can be driven into the reference plane (see Figure 3-5) determined by the two horizontal axes a and b at right angles. For that purpose, the base B is shifted by z_0. The sample surface can further be adjusted with respect to the reference plane by a tilt correction of Θ and χ. Each spot of the sample can be placed into the measuring position or tilt center T by a lateral movement x and a rotation ϕ on the vertical axis. Finally, the angle of incidence can be varied by a tilt Θ around the axis a lying on the wafer surface.

All six movements can be driven by stepping motors. The z-shift should have a range of 10 mm with a step size of 1 μm. The tilts Θ and χ should be variable between $0°$ and $3°$ at a step width of $0.001°$. Coarse steps are satisfactory for the movement x and rotation ϕ, but a large range of adjustment is necessary in this case (> 100 mm for x and $> 180°$ for ϕ). The height z_0 of the base has to be corrected very rarely, and this can be done manually.

The adjustment of a sample for thin-layer analysis is obviously not as easy as fixed-angle positioning for micro- or trace analysis. The angle of incidence has to be set with an absolute *accuracy* of $< 0.005°$ and has to be varied in a range of about $1°$. The stepwise variation is needed to record an angle-dependent intensity profile, which is described in detail in Section 4.4.1.

3.5. DETECTION AND REGISTRATION

The fluorescence radiation of the sample has to be detected and registrated as an X-ray spectrum. This problem can generally be solved either by a wavelength-dispersive or an energy-dispersive spectrometer (WDS or EDS). For TXRF instruments, only the EDS has been applied so far. While a WDS requires a goniometer, an EDS needs no mechanically moving parts. It simply uses an electronic system based on a special semiconductor detector. There are no problems of mechanical alignment or replacement apart from the necessity of providing the detector with liquid nitrogen (about 10 L/week). But today's detectors must not be cooled permanently. They can be warmed up and returned to the temperature of liquid nitrogen (77 K or $-196°C$) repeatedly. At present detectors that are cooled by a Peltier element or are operating at room temperature have rather poor resolution, but there may well be a technological breakthrough in the future.

The complete spectra of an EDS are registered simultaneously and not sequentially. Thus, the time-consuming mechanical scan of a WDS is avoided. Furthermore, the semiconductor detector of an EDS can be placed very close to the sample, thereby receiving a wide cone of fluorescence radiation. The considerably increased intensity is used to substantially reduce the counting time. An EDS can record a total X-ray spectrum on the order of seconds instead of minutes, as is necessary for a WDS.

Because of the simplicity, speed, and convenience of operation, the EDS is adapted to all TXRF instruments without exception. Some essential limitations as regards its spectral efficiency must therefore be accepted. The main disadvantage of an EDS is poorer spectral resolution for photon energies below 15 keV. This causes a troublesome peak overlap, especially below 3 keV, although interferences with high-order reflections appearing in WDS do not occur in EDS systems. Furthermore, the efficiency of an EDS strongly diminishes for photon energies below 2 keV, leading to a weaker detection of lighter elements. Nevertheless, an EDS is a convenient and economical system for complex multielement analyses.

In such an EDS, the X-ray photons emitted from the sample are directly collected by a semiconductor detector. This special solid-state detector does not merely count the individual photons but can also determine their different energies. For any collected photon, it produces an individual voltage pulse the amplitude of which is proportional to the energy of this photon. All the detector pulses are processed in an electronic measuring chain and finally sorted by a multichannel analyzer. The content of this counter storage can be represented as the particular X-ray spectrum and can be further processed directly by a dedicated computer.

Reflectivity measurements may be taken by an additional detector that can simply be incorporated into the arrangement just described [20, 21]. The primary beam monochromatized by a first reflector is reflected at the layered or unlayered substrate S. The intensity of this reflected beam can be measured by a second simpler detector. This may be a photodiode or a scintillation detector—much simpler and cheaper than a Si(Li) detector used for TXRF. But it must be ensured that this second detector is tilted at the double angle 2α when the layered sample is tilted at the single angle α—both around the same axis a of Figure 3-8.

3.5.1. The Semiconductor Detector

The heart of any EDS is a special solid-state—or rather a semiconductor— detector [22–24]. It basically consists of a pure silicon or germanium crystal. This crystal ought to be several millimeters wide and thick and should be extremely resistive. Germanium can be purified by zone refining to achieve the necessary ohmic resistance. But silicon cannot be produced with such a high degree of purity [impurities of several parts per billion (ppb)]. The most common impurity is the element boron, which modifies silicon to a p-type semiconductor with decreased resistivity and increased conductivity. In order to suppress this effect, another impurity is artificially added to the crystal. Usually, the boron "acceptors" are compensated or neutralized by lithium "donors." The lithium diffuses into the crystal at elevated temperature and "drifts" under the influence of an electric field. In this way, a crystal with a high intrinsic resistivity is produced with a thin p-type layer and n-type layer at the end planes and a large intrinsic region between them. Such a crystal is termed a lithium-drifted silicon crystal, or a Si(Li). There are also lithium-drifted germanium crystals, or Ge(Li)'s. But these are more and more being replaced by high-purity germanium (HPGe).

As illustrated in Figure 3-9, the frontal areas of the crystal are coated with thin layers of gold serving as electrodes. An inverse dc voltage is applied called a *reverse bias* (p-type layer negative; n-type layer grounded). It defines the direction of low conductivity, i.e., of a small leakage current in spite of the high voltage (-500 to -1000 V). The total configuration is termed a p-i-n diode with a reverse bias. It is cooled to the temperature of liquid nitrogen (77 K) for two reasons: (i) to reduce the thermal leakage current even further, and (ii) to prevent a reverse diffusion of the lithium ions, i.e., to freeze-in the compensation state. The second reason does not apply to HPGe when it is used as the intrinsic material. But the first reason requires that the germanium crystal be cooled as well.

The mode of operation is the same for both detectors: the Si(Li) detector and the HPGe detector. (It is illustrated in Figure 3-9). An incident X-ray photon interacts with the crystal and ionizes the crystal atoms creating

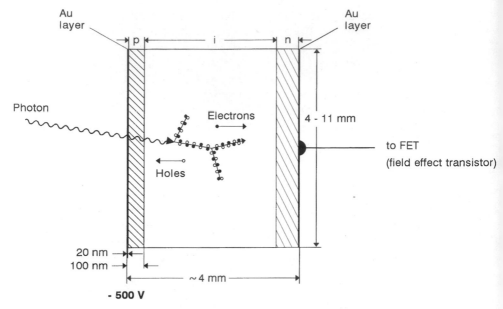

Figure 3-9. A semiconductor detector operated as a *p-i-n* diode with a reverse voltage or bias. An incident X-ray photon ultimately produces a series of electron–hole pairs. They are "swept out" by the bias field of $-500\,V$: electrons in the direction of the *n*-layer; holes in the direction of the *p*-layer. Thus, a small charge pulse is produced.

photoelectrons and Auger electrons. These electrons then pass on their energy in several steps and raise outer electrons from the valence band into the conduction band of the crystal lattice. Simultaneously, electron holes are created in the valence band. A total track of electron–hole pairs is produced until the energy of the incident photon is used up. Because of the applied high voltage, the electron–hole pairs separate and the electrons and electron holes rapidly drift to the positive electrode and the negative electrode, respectively. A charge pulse is produced usable for single-photon counting. Since the number of electron–hole pairs is directly proportional to the energy of the particular photon detected, the magnitude of the charge pulse is proportional to the photon energy as well. Consequently, the charge pulse gives a measure of the energy of the detected photon. Hence, the semiconductor detector is capable of counting single X-ray photons and of reading their different energies as well. This feature is the prerequisite for an EDS.

Figure 3-10 shows a sectional view of the front end of a Si(Li) detector. The grooved silicon crystal is coated with gold layers serving as electrodes for the high voltage. It is connected to a copper rod immersed in liquid nitrogen. A field-effect transistor (FET) used as a preamplifier is installed between the

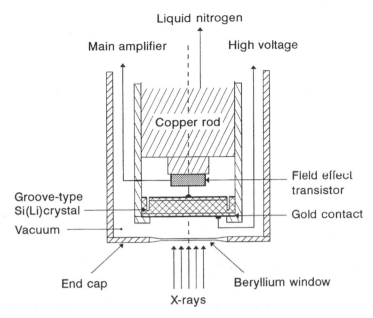

Figure 3-10. Cross section of the front end of a solid-state or semiconductor detector, here with a grooved Si(Li) crystal. Crystal and preamplifier are connected with a cooled copper rod and shielded by a case with an end cap and Be-window. (After Williams [24] and Ellis [25].)

detector and the cooled copper rod. The inner device is encapsulated in an outer metallic case locked by a thin beryllium window. This case is evacuated for good thermal insulation of the cooled inner device. The consumption of liquid nitrogen is kept low by this means, and the window does not become covered with moisture. The beryllium window itself can be traversed by X-rays with a fairly low attenuation, and it protects the crystal against air, dust, moisture, and light.

3.5.2.   The Registration Unit

The different charge pulses produced by the detector are processed by an elaborate electronic system. They are amplified, shaped, and sorted according to their amplitudes. Finally, all pulses with certain amplitudes are counted. A strong proportionality between the amplitude of the initially produced pulses and subsequently processed pulses should accurately be maintained. As long as this condition is met, the pulse amplitude remains an accurate measure of the energy of the detected photons. Ultimately, the result of the photon counting can be demonstrated as an energy-dispersive spectrum representing the number of photons as a function of their energy.

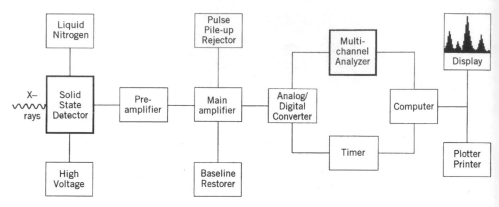

Figure 3-11. Major components of an energy-dispersive spectrometer with a solid-state detector and a multichannel analyzer.

The electronic systems employed usually have the components shown in the block diagram of Figure 3-11: an FET-based preamplifier, a main linear amplifier with baseline restorer and pulse-pileup rejector, an analog-to-digital converter (ADC), a multichannel analyzer (MCA), a timer, and a dedicated computer with a color video display and a plotter/printer output device.

An FET is the most used preamplifier operated in the mode of pulsed optical feedback for discharge. Since around 1988, a Pentafet[1] has been used for direct charge neutralization [25, 26]. The preamplifier converts the charge pulses produced by the detector to low-voltage pulses. The amplitudes or pulse heights are strictly proportional to the number of electron–hole pairs and hence to the energies of detected X-ray photons. The preamplifier is installed quite close to the detector and also cooled to the temperature of liquid nitrogen in order to reduce the electronic noise.

The millivolt pulses are further increased to voltage levels by a high-gain linear amplifier. In addition, the pulses are shaped to a particular form with different amplitudes but with a constant shaping time. Effective noise suppression is achieved by selecting a sufficiently high shaping time (6 or 8 μs rather than 1 or 2 μs). On the other hand, the probability of a pulse overlap increases with higher shaping time. Such an overlap of two coinciding pulses leads to the registration of one pulse the amplitude of which is the sum of the individual pulse amplitudes. This incorrect registration is called the *pulse-pileup effect*. It can be avoided by a device called a *pulse-pileup rejector* at least in those cases when the two pulses are not exactly coincident in time but arrive at least 1 μs apart. These nearly coincident pulses are "rejected" and get lost in

---

[1] Registered Trademark of Oxford Instruments, High Wycombe, England.

the counting. But the loss can be compensated by an appropriate extension of the measuring time.

The pileup effect becomes serious at high count rates. In such cases, not only coincident but also successive pulses will overlap at least with their "tails." Consequently, the baseline of the amplifier is shifted to a higher voltage and the output pulses appear reduced or "depressed" below their true amplitude. This incorrect reading can be avoided electronically by another device called a *baseline restorer*.

The output pulses of the main amplifier are transferred to an ADC. The analog information of the pulse amplitudes is converted here into digital form. The resultant number serves as an address for the connected MCA. It sorts the different pulses according to their address and counts the number of respective pulses. The data are stored in the MCA memory, which is similar to that of a conventional computer. The different "bins" or "channels" correspond to small energy ranges. Usually, an MCA contains 1000 channels, which are assigned to consecutive increments of 10, 20, or 40 eV. Hence, a total range of 10, 20, or 40 keV is indicated for the energy of X-ray photons. The storage capacity usually amounts to 2^{16}, i.e., 65,536 counts per channel (16 bits) or even to 2^{32}, i.e., 4×10^9 counts per channel (32 bits; hence called double precision).

The digital form of an MCA's contents is well adapted to computer operation. The raw data can easily be processed by a dedicated minicomputer or a personal computer (PC). The relevant contents can be shown as an energy-dispersive spectrum and can either be displayed on a monitor or plotted on a printer/plotter. The spectrum is usually represented as a histogram indicating counts vs. energy and can already be observed during the measurement. The computer can control the output devices and also a device called a *timer*, which is used to start and to stop the measurement or data collection at a preset time. In most systems, the timer is based on a "live-time" clock. It stops any further pulse counting during the dead time of the system when an input pulse is still being processed.

3.5.3. Performance Characteristics

The performance of an EDS with semiconductor detector and electronic pulse processor is characterized by four main features: the spectral efficiency of the detector, the spectral resolution of the system, the input–output yield of the processor; and the troublesome escape-peak phenomenon caused by the detector.

3.5.3.1. Detector Efficiency

The efficiency ε of a semiconductor detector is defined as the percentage of detected photons with respect to the incident photons. The efficiency is nearly

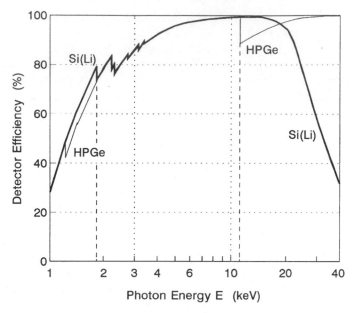

Figure 3-12. Efficiency of typical Si(Li) (———) and HPGe (———) detectors calculated as dependent on the energy of the indicated photons. The Si(Li) detector was assumed to have a 3 mm thick Si-intrinsic region, the HPGe detector to have a 5 mm pure Ge-crystal. Both detectors are assumed to be provided with a 20 nm thick gold contact and a 7.5 μm thick Be-window.

100% for photons with an energy between 6 and 11 keV but is reduced for lower and higher energies, as is shown in Figure 3-12. The lower efficiency is caused by a reduced transmission of X-rays in the beryllium window, in the gold layer, and in an inactive or dead layer of the Si(Li) crystal. Furthermore, it is caused by a limited photoelectric absorption of X-rays in the intrinsic region of the Si(Li). These effects can be described by the following formula [27]:

$$\varepsilon = \exp\left[-(\mu/\rho)_{Be}\,\rho_{Be}\,t_{Be} - (\mu/\rho)_{Au}\,\rho_{Au}\,t_{Au} - (\mu/\rho)_{Si}\,\rho_{Si}\,t_{Si}\right]$$
$$\cdot\left\{1 - \exp\left[-(\tau/\rho)_{Si}\,\rho_{Si}\,d_{Si}\right]\right\} \tag{3-11}$$

where (μ/ρ) is the mass-attenuation coefficient; ρ, the density; t, the thickness of the particular layer; and d, the thickness of the detector crystal. This formula is written for a Si(Li) detector. For a HPGe detector, the silicon values have to be replaced by germanium values.

The individual data of typical Si(Li) and HPGe detectors are listed in Table 3-3 and used as the basis for Figure 3-12. The calculated curves only serve for a general assessment. If true values of detector efficiency are required, an experimental determination should be carried out.

Table 3-3. Data for the Absorbing Components of Typical Si(Li) or HPGe Detectors

Absorbing Layer or Medium	Density ρ (g/cm^3)	Typical Thickness (nm or μm or mm)
Beryllium window	1.85	7.5 μm
Gold contact	19.3	20 nm
Silicon dead layer	2.33	0.1 μm
Silicon intrinsic region	2.33	3 mm
Germanium dead layer	5.32	1 μm
Highly pure germanium	5.32	5 mm

At photon energies below 2 keV, the absorption of the Be-window is the decisive factor. At photon energies above 20 keV, the transmission through the small intrinsic or active Si crystal becomes dominant. Different steps of the efficiency curves can be identified: five small steps at the M-edges of Au between 2.2 and 3.4 keV; one small step at the K-absorption edge of Si at 1.838 keV; one larger step at the K-edge of Ge at 11.103 keV; and a smaller one at the LIII-edge at 1.217 keV.

As shown in Figure 3-12, both detectors differ strongly in the region above 11 keV. After a stepwise decrease, the Ge detector attains 100% efficiency for energies above 20 keV whereas the Si(Li) detector steadily looses efficiency. In the region below 11 keV, the detectors have similar characteristic curves. Their low efficiency at photon energies below 2 keV is diminished still further by an air path between the sample and the detector. For a typical distance of about 5 mm, the overall efficiency is reduced to about 65% for 2 keV photons and to about 2% for 1 keV photons.

3.5.3.2. Spectral Resolution

The different characteristic X-ray peaks of an EDS cover an energy range of some 100 eV. The histograms usually recorded show peaks that span about 10–20 channels, with nearly a Poisson or Gaussian distribution. Each peak can be characterized by a width defined as the full width at half-maximum (FWHM).

The peak width is mainly caused by the production of electron–hole pairs by the incident photons. But this process is not the only one possible; rather, it competes with, e.g., the generation of lattice vibrations. Without such a competing process, the necessary energy for the generation of an electron–hole pair would amount to the band-gap energy of 1.1 eV for silicon and 0.7 eV for germanium. Because of the competing processes, however, the average energy consumed per electron–hole pair is greatly increased: 3.85 eV for silicon and 2.95 eV for germanium at the operating temperature of 77 K [22]. Moreover, the number of electron–hole pairs is no longer constant but fluctuates around a statistical mean. A corresponding frequency distribution may be found for the pulse amplitudes, which leads to the observed peak width.

Such a distribution would be a Poisson distribution if independence and a small probability for the observed events could be taken for granted. However, the processes responsible occur fairly frequently, rather than rarely, and not independently. Consequently the distribution is not a true Poisson distribution [24], but it can be approximated by what is called a quasi-Poisson distribution. The deviation can be characterized by the Fano factor F. It also determines the significantly smaller peak width ΔE in the spectrum, measured by the FWHM:

$$\Delta E = 2.35 \sqrt{FeE} \qquad (3\text{-}12)$$

where e is the average energy per electron–hole pair, and E is the energy of the detected photons.

Unfortunately, there is a second independent source of fluctuations. They arise from the electronic noise of the preamplifier in spite of low-temperature cooling and short connecting cable and lead to an additional spread of the peaks. The total width ΔE becomes [22, 24]

$$\Delta E = \sqrt{\Delta E_{noise}^2 + 5.55\, FeE} \qquad (3\text{-}13)$$

where ΔE_{noise} is the additional component of the preamplifier, independent of the peak energy E. Even though it is quite small, ΔE_{noise} represents the lower limit of the spectral resolution of an EDS.

The peak width ΔE, which may be called the spectral resolution, is graphically presented in Figure 3-13 as a function of the photon energy E. The Fano factor F was chosen to be 0.11 and the noise value ΔE_{noise} to be 75 eV for a detector crystal of 50 mm². These values are typical for good-quality detectors currently available. In the figure, note that the HPGe detector exhibits better resolution than the Si(Li) detector but the differences are below the 10% level. Both detectors are much poorer in resolution than a WDS with LiF or PET (pentaerythritol) crystals used as Bragg reflectors; however, their resolution is amply sufficient to separate the $K\alpha$-peaks of neighboring elements in the periodic system.

Conventionally, the spectral resolution is only specified by the Mn-$K\alpha$ peak with a photon energy of 5.9 keV. This is a simplification of the more complex relationship represented by Figure 3-13. But the single value of the peak width of Mn-$K\alpha$ is well suited to characterize the influence of the size of the detector crystal. The size of the frontal area determines its capacitance and influences

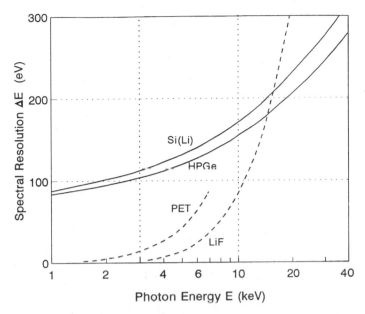

Figure 3-13. Spectral resolution or peak width ΔE as a function of photon energy or peak position E. The resolution was calculated for high-quality detectors with a frontal area $A = 50\,\text{mm}^2$, a noise-component $\Delta E = 75\,\text{eV}$, and a Fano factor $F = 0.11$. The resolution for the two detectors in EDS is compared to that of LiF or PET crystals in WDS.

Table 3-4. Influence of the Crystal Size of Si(Li) and HPGe Detectors on the Spectral Resolution ΔE of the Mn-$K\alpha$ Peak

Size of Detector Crystal		Spectral Resolution		
Diameter (mm)	Frontal Area (mm^2)	Noise ΔE_{noise} (eV)	Si(Li) ΔE (eV)	HPGe ΔE (eV)
3.6	10	50	128	115
6.2	30	62	133	120
8.0	50	75	140	127
10.1	80	95	151	140

Source: After Ellis [25].

the noise component ΔE_{noise} [25]. Table 3-4 presents a survey of crystals of different sizes and their influence on the resolution for Si(Li) and HPGe detectors.

The spectral resolution is usually determined at a low photon flux or count rate below 1000 cps. But the resolution of an EDS is also dependent on this flux of the incident X-ray photons. The next subsection considers this dependence.

3.5.3.3. Input–Output Yield

The number of pulses produced by the detector crystal within a certain time interval is called the *input rate*. The number of pulses being processed by the electronic chain and indicated by the MCA is defined as the *output rate*. The input and output rates differ from one another due to the dead-time effect of the electronic system.

The pulses coming from the detector have a certain shaping time that can be chosen between 1 and 8 μs. They are transmitted to the MCA if they do not overlap within a period of 5–40 μs. Otherwise, their registration is refused by the pulse-pileup rejector. This period of rejection is the governing dead time of the system. Because of this dead time, the indicated output rate is always smaller than the given input rate. It falls more and more behind an increasing input rate due to increasing dead-time losses.

Characteristic input–output curves are presented in Figure 3-14 in a double-logarithmic plot. Four different sets are chosen here for the shaping time and the dead time. They produce differences in the spectral resolution serving as the parameter of the curves, i.e., the width ΔE of the Mn-$K\alpha$ peak assigned to the four different curves.

For small input rates, the output rate is at first linearly increasing. But then this increase slows down. After reaching a maximum, the output rate even decreases. The *maximum* output rate is inversely proportional to the shaping time. With a shorter shaping time, the dead time can be reduced and a higher count rate can be registered. On the other hand, the spectral resolution of the detection system deteriorates by increased electronic noise that leads to widening of the peaks. A reasonable compromise has of course to be made between a high count rate and a small peak width, i.e., a "good" spectral resolution.

The deviation from a linear relationship between input and output rate is caused by dead-time losses. If n is the input rate and τ is the decisive dead time, the system will be "dead" during a percentage D of the total time given by

$$D = n \cdot \tau \cdot 100\% \qquad (3\text{-}14)$$

and it is "live" for the remaining $(100 - D)$ in percent. At the same time, D means the percentage of dead-time losses. Output rates with a constant D-value are represented by straight (dashed) lines in the double-logarithmic plot of Figure 3-14. They are parallel to the *ideal* straight line for which the output and input rates are equal, i.e., $D = 0\%$.

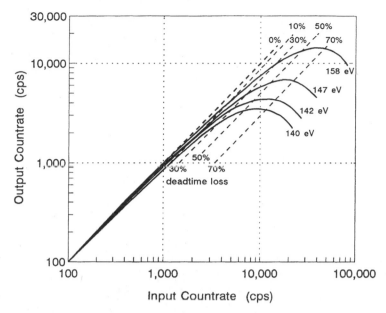

Figure 3-14. Indicated output rate dependent on the true input rate of an EDS system with a Si(Li) detector of 50 mm² front area. Due to the dead-time losses, the curves deviate from the ideal straight line and finally show a choking of the detector. Each curve corresponds to a certain spectral energy resolution ΔE (in eV). Constant dead-time losses are indicated by dashed straight lines with the D-values shown. (After Woldseth [22] and an Oxford Instruments technical booklet [26].)

To determine the actual input rate from the registered output rate, the dead-time losses have to be corrected for. Up to the respective maximum, the actual input rate can be read unambiguously from the curves of Figure 3-14. Beyond this maximum, however, a measured output rate would lead to a second value for the unknown input rate. In order to avoid this ambiguity, output registration and input reading have to be restricted to values below the maximum.

In practice, the dead-time losses are compensated by a prolongation of the measurement. This can be achieved by stopping the timer for the period of dead time. If a "live" time t is preset for the measurement, the actual time for the acquisition of pulses will be extended to an "acquire" time T. For a given dead-time loss D, it must be equal to

$$T = \frac{100}{100 - D} t \qquad (3\text{-}15)$$

If the D-value amounts to 50%, for instance, the acquire time is extended to the double of the preset live time. For this case, the input rate is just double the output rate.

The dead-time losses D can be indicated by a special loss meter. As can be seen from Figure 3-14, the 50% level roughly determines the maximum of the output rate. This limit should not be exceeded, as mentioned earlier, but for a high count rate the dead-time loss should come close to this value. This rule can be a guideline for setting up excitation parameters and for achieving the optimum performance of an EDS.

3.5.3.4.   The Escape-Peak Phenomenon

A peculiar phenomenon occurs with the appearance of spurious peaks, called *escape peaks*, in energy-dispersive spectra. They can be a real nuisance when they coincide with the small peaks of trace elements.

Escape peaks arise when a strong element peak is recorded by the detector. Accordingly, they can be regarded as "daughter" peaks produced by a strong "mother" peak. Their formation occurs within the detector crystal. When an incident X-ray photon is passing through the crystal and its characteristic energy E is sufficiently high, it can produce a photoelectron from an inner shell of a crystal atom. As a result, the excited atom can emit an X-ray photon by fluorescence, mostly a $K\alpha$ or $K\beta$ photon. It is normally reabsorbed in the crystal, creating a chain of electron–hole pairs and thus contributing to the charge pulse of the detector. However, this $K\alpha$ or $K\beta$ photon can also escape from the crystal with a certain probability. In that case, it carries off the quite definite energy $E_{K\alpha}$ or $E_{K\beta}$ of the element the crystal is composed of and does not produce its own chain of electron–hole pairs. On the one hand, this photon is lost for purposes of detection and so the detector efficiency is subject to the discontinuities already shown in Figure 3-12. On the other hand, the residual energy shows up as an individual photon of the actual energy $E - E_{K\alpha}$ or $E - E_{K\beta}$. Such "packages" of energy or "photons" appear as separate peaks in the spectrum. Figure 3-15 shows a few such escape peaks due to some strong mother peaks. As is shown in the figure, their appearance is quite different for a Si(Li) detector and for a HPGe detector.

An effect similar to the escape-peak phenomenon will occur if secondary electrons instead of X-ray photons escape from the detector volume, mainly from a near-surface layer. This effect is called *incomplete charge collection*. It leads to a reduction and tailing of the mother peak. The spectral background on the low-energy side of the mother peak is thereby lifted, but only slightly ($< 0.1\%$ of the peak height in a distance > 500 eV).

The position of escape peaks is dependent on the position of the mother peak. Their peak height is mainly dependent on the fluorescence yield and the

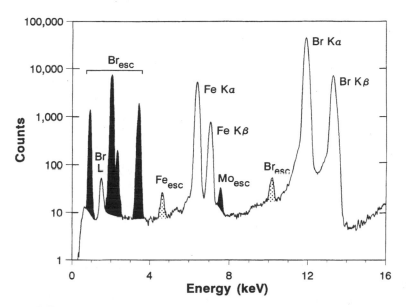

Figure 3-15. Energy-dispersive spectrum of $FeBr_2$ excited by a Mo X-ray tube and recorded by a Si(Li) detector (semilogarithmic). The Si-escape peaks of Fe and Br are filled with dots. If a HPGe detector is used instead of the Si(Li) detector, four escape peaks of Br and even one of Mo will distinctly appear whereas the two Si escape peaks will vanish. The Ge escape peaks are marked in black.

mass-absorption coefficient of the detector crystal. This is demonstrated in Figure 3-16 for a Si(Li) detector and a HPGe detector. Escape peaks will only arise if the mother peak lies "above" the energy of the respective absorption edge of the crystal (1.838 keV for Si; 11.103 keV for Ge). The position of the escape peak is given by

$$E_{esc} = E_{mother} - E_{K\alpha} \qquad \text{or} \qquad E_{esc} = E_{mother} - E_{K\beta} \qquad (3\text{-}16)$$

where $E_{K\alpha}$ is 1.740 keV for silicon and 9.876 keV for germanium; $E_{K\beta}$ is 1.830 keV for silicon and 10.984 keV for germanium. The Si escape peaks are much smaller than the Ge escape peaks. For Si(Li) detectors, the escape peaks attain a height of only about 1% of the mother peak and each $K\beta$ escape peak is strongly overlapped by the respective $K\alpha$ escape peak. For Ge detectors, not only the $K\alpha$ escape peak but also the relevant $K\beta$ escape peak can separately be observed. Their peak heights reach up to 20% of the mother peak, leading to several troublesome interferences.

The total effect is more of a problem during the identification of small trace-element peaks overlapped by escape peaks. It causes fewer problems as

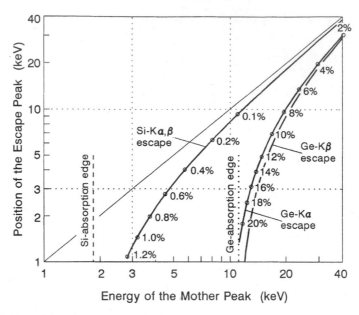

Figure 3-16. Position of *escape* peaks of a Si(Li) detector and a HPGe detector dependent on the energy of their *mother* peak. The percentage values noted on the curves give the peak height of the escape peaks in relation to that of the mother peaks. (After Woldseth [22].)

regards counting losses of a strong mother peak. Also, it is less troublesome for Si(Li) detectors than for Ge detectors, where it poses difficulties throughout the important region of X-ray spectra between 1 and 30 keV. But these problems disappear if the sample consists only of elements with mother peaks below 11 keV, e.g., if only lighter elements ($Z < 31$) are present.

REFERENCES

1. Marten, R., Rosomm, H., and Schwenke, H. (1977). German patent pending, No. P 26 32 001.4.

2. Iida, A., Yoshinaga, A., Sakurai, K., and Gohshi, Y. (1986). *Anal. Chem.* **58**, 394.

3. Rieder, R., Wobrauschek, P., Ladisich, W., Streli, C., Aiginger, H., Garbe, S., Gaul, G., Knöchel, A., and Lechtenberg, F. (1995). *Nucl. Instrum. Methods* **A355**, 648.

4. Iida, A., Gohshi, Y., and Matsushita, T. (1985). *Adv. X-ray Anal.* **28**, 61.

5. Rich. Seifert & Co. (1985). *X-ray Tubes*, technical booklet. Seifert, Ahrensburg, Germany.

6. Rigaku International Corporation (1992). *Rotating Anode X-Ray Generator Systems*, technical brochure. Rigaku Europe GmbH, Düsseldorf, Germany.

7. Schwenke, H., Knoth, J., Marten, R., and Rosomm, H. (1978). German patent pending, No. P 27 36 960.4.

8. Knoth, J., Schneider, H., and Schwenke, H. (1994). *X-Ray Spectrom.* **23**, 261.

9. Ladisich, W., Rieder, R., Wobrauschek, P., and Aiginger, H. (1993). *Nucl. Instrum. Methods* **A330**, 501.

10. Schwenke, H., and Knoth, J. (1982). *Nucl. Instrum. Methods* **193**, 239.

11. Kregsamer, P., and Wobrauschek, P. (1991). *Spectrochim. Acta* **46B**, 1361.

12. Wobrauschek, P., Kregsamer, P., Ladisich, W., Rieder, R, and Streli, C. (1993). *Spectrochim. Acta* **48B**, 143.

13. Schwenke, H., and Knoth, J. (1993). "Total Reflection XRF." In *Handbook of X-Ray Spectrometry* (R. van Grieken and A. Markowicz, eds.), Practical Spectroscopy Series, Vol. 14, p. 453, Dekker, New York.

14. Schuster, M. (1991). *Spectrochim. Acta* **46B**, 1341.

15. Schmitt, M., Hoffmann, P., and Lieser, K.-H. (1987). *Fresenius' Z. Anal. Chem.* **328**, 594.

16. Kollotzek, D. (1980). Diploma thesis, University of Stuttgart.

17. Prange, A., Kramer, K., and Reus, U. (1993). *Spectrochim. Acta* **48B**, 153.

18. Wobrauschek, P., and Aiginger, H. (1979). *X Ray Spectrom.* **8**, 57.

19. Prange, A., and Schwenke, H. (1992). *Adv. X-Ray Anal.* **35B**, 899.

20. Philips Analytical (1991). *Characterization of Thin-Layered Samples by Glancing Incidence X-ray Analysis*, technical brochure. Philips, Eindhoven, The Netherlands.

21. Hüppauf, M. (1993), Ph.D. thesis, RWTH Aachen; Berichte des Forschungszentrums Jülich, Jül-2730.

22. Woldseth, R. (1973). *X-Ray Energy Spectrometry*. Kevex Corporation, Burlingame, California.

23. Thompson, A.C. (1986). *X-Ray Data Booklet*. Lawrence Berkeley Laboratory, Berkeley. California, pp. 6–7.

24. Williams, K.L. (1987). *An Introduction to X-Ray Spectrometry*. Allen & Unwin, London.

25. Ellis, A.T. (1991). *PICXAM-Workshop*, Hilo, Hawaii. Instruction material.

26. Oxford Instruments (1993). *Link EDX Detectors*, technical booklet. Oxford Instruments, High Wycombe, England.

27. Klockenkämper, R., and von Bohlen, A. (1989). *Spectrochim. Acta* **44B**, 461.

CHAPTER

4

PERFORMANCE OF TXRF ANALYSES

An analytical strategy has to be based on the prerequisites of the method to be applied. Above all, TXRF is restricted to small sample amounts. Only micrograms of a solid material and less than 100 μL of a liquid can be analyzed in one go. Consequently, TXRF is a method of microanalysis, as defined by IUPAC [1], and samples can seldom be analyzed as received. A certain pretreatment is generally required, in contrast to conventional XRF. Samples have to be prepared as solutions, suspensions, fine powders, or thin sections. Solids must be ground or dissolved. For a determination of ultratraces, the matrix of the sample should first be separated and removed. For that purpose, all techniques can be used that have already been tested and combined with other methods of atomic spectroscopy, e.g., AAS or ICP-OES. Certain precautions have to be taken in dealing with small samples, and a clean-bench working procedure is mandatory for critical steps of sample preparation.

On the other hand, TXRF is a variant of energy-dispersive X-ray spectrometry and shares all the convenient features. The complete spectrum is recorded simultaneously within seconds; it is displayed on a screen, and the registration can be observed continually during the measurement. A dedicated computer is usually incorporated for advanced processing of the spectra. Automatic peak identification is made possible, enhancing the speed and ease of qualitative analysis. A visual comparison of two complete spectra enables a fingerprint analysis.

Quantitative analysis by TXRF is essentially facilitated by the use of only small amounts of sample. Troublesome matrix effects do not arise— neither absorption nor enhancement effects. Quantification can therefore be carried out after the addition of an element serving as the internal standard. For a single-element analysis, the analyte can itself be used as the standard. For multielement analyses, any element not present in the sample can be chosen as the standard against which all the other elements are to be determined. In this case, the different sensitivity values for these elements are needed. They have to be determined prior to analysis, but only once for each new instrument.

Surface and thin-layer analyses are carried out by varying the glancing angle in the region of total reflection. At the same time, the peak intensity of concerned elements is recorded. These angle-dependent intensity profiles give a first qualitative picture of contaminants, layers, and/or a substrate. But quantification by an internal standard is not possible since the primary beam not only passes through a thin upmost layer but also penetrates to a greater depth when the critical angle of total reflection is exceeded. To get a quantitative description of the layered system, an algorithm has to be applied that is already known from conventional XRF and is called the *fundamental-parameter method*. It is based on a simple model in order to calculate fluorescence intensities of the individual elements while allowing for matrix effects. Only one external standard is needed. The fundamental data can be obtained from tables or partly calculated by use of equations.

4.1. PREPARATIONS FOR MEASUREMENT

First of all, TXRF is a method of microanalysis. It is directly applicable when only a small sample amount is available or when only a small part of a larger amount can be taken. But when a large sample amount is received for analysis, an appropriate sampling of a smaller part has first to be carried out. This part may be called the *specimen* is contrast to the total amount, which is called the *sample* [2]. In order to get a representative result by means of a specimen, the sample has to be homogeneous from the beginning or thoroughly homogenized prior to sampling. At the same time, the main constituent or the matrix of the sample should be removed if possible so that the analyte is enriched and essentially present in the actual specimen. Finally, this specimen is placed on a flat carrier for analysis. All these steps of preparation and presentation have to be carried out very carefully. Only clean or specially cleaned vessels, instruments, and of course carriers should be employed. In general, only analytically pure ("p.a." = pro analysi) or suprapure reagents may be used. Acids should be finally purified by subboiling; water should be prepared by a double-stage deionization or bidistillation.

Highly pure acids needed for sample preparation are commercially available in the p.a. grade or suprapure. They may have, however, some residual impurities on the order of ng/mL. Especially elements like Mg, Al, Fe, Cu, Zn, and Pb may show high blank values. The same problem arises for water even if it is deionized or bidistilled: Cl, K, Ca, Br, and Sr are the troublesome elements here. Purification of these liquids is however possible by *subboiling* distillation. The liquid is vaporized by heating below the boiling point. The vapor is condensed at a cooling finger, and the purified condensate is collected in a small flask. Impurities are significantly below 1 ng/mL.

4.1.1. Cleaning Procedures

For the preparation of samples, only vessels made of quartz glass (Suprasil or Synsil) PTFE (polytetrafluorethylenes, e.g., Teflon), or PP (polypropylenes) should be used. Due to their high purity, contamination by these materials is greatly reduced. Also, vessels made of PFA (perfluoroalkoxy polymers) can be recommended because of their especially smooth walls. Before the vessels are used, they must be cleaned by boiling and additional steaming (except for PP). Figure 4-1 represents a *steaming* device as it is usually employed [3]. Pure nitric acid and pure water are recommended as cleansing agents and should be used successively. The liquids are vaporized, and the respective extremely

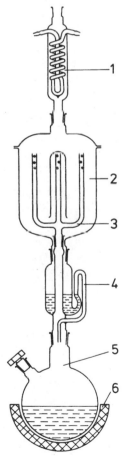

Figure 4-1. Setup of a steaming device used for a final cleaning of glass vessels and instruments: 1, condenser; 2, steam chamber; 3, rack of quartz glass as a support for vessels; 4, overflow; 5, glass flask; 6, heating jacket. (After Tschöpel et al. [3].)

clean vapor is condensed at the cool vessels in the steaming chamber. The condensate takes up the impurities and moves them into the liquid phase. This process is continued for a period of up to 12 h.

The specimen should be placed on highly clean carriers. Plexiglas carriers can be applied without cleaning. These cheap carriers are used only once. All the other more expensive carriers are used frequently for cost saving. They must be cleaned even before their first use because they are not delivered in sufficiently clean condition.

A simple *cleaning* process can be recommended that works even without a time consumptive steaming. This process is preferably carried out for a total set of several carriers. Figure 4-2 shows a special support made of PTFE and designed to take 24 different carriers with a diameter of 30 mm. This support is immersed in a large beaker of 800 mL, half filled with a cleaning agent (e.g., RBS 50 in a 10% dilution; Carl Roth GmbH & Co., Karlsruhe, Germany), which is brought to a boil and cooled down. The support with the carriers is rinsed with distilled water and put into a second beaker with Milli-Q water (Millipore Corp., Bedford, Massachusetts). This water is boiled and then

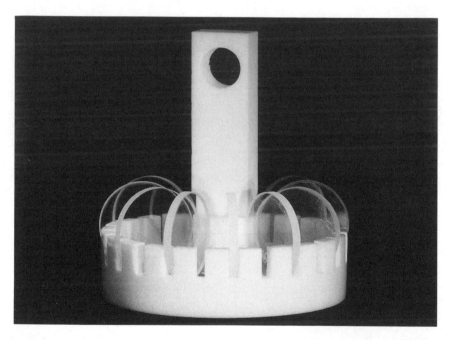

Figure 4-2. A special support for carriers used in the cleaning process. The support is made of Teflon and is designed for 24 carriers. (Available from Atomika Instruments, GmbH, Oberschleissheim, Germany.)

cooled down to about 40 °C, after which all carriers and the support are dried up with fluff-free precision wipers (Kimwipes, Kimberly-Clark Corp., Northop, UK). Thereafter, the support with the carriers is placed in a third beaker with concentrated nitric acid (p.a. grade, E. Merck, Darmstadt, Germany), boiled for 1 h, cooled down, and again placed in the second beaker with fresh ultrapure water. It is warmed up to about 60 °C for 1 h. Each bath should be prepared freshly, and the beakers must be covered appropriately. The boiling must take place in a clean fume cupboard only used for such a purpose in order to preclude any source of contamination. The support and carriers must not be touched by hand.

After a last cooling to 40 °C in a clean bench, the support with the carriers is lifted out and remaining droplets are wiped off. The dried carriers are put into clean Petri dishes. Covered by the top, they are kept in a drawer until needed for analysis. The total procedure takes about 4 h. The clean carriers can be used for deposition of solid samples without restrictions. For deposition of liquid samples they must be hydrophobic, or otherwise the deposited droplets run out. Silicon, glassy carbon, boron nitride, and Plexiglas are hydrophobic by nature but not quartz glass. It is initially to be coated with a hydrophobic film. For that purpose, a $2 \mu L$ droplet of a silicone solution (Serva GmbH & Co., Heidelberg, Germany) is usually pipetted onto the quartz glass carriers. It is spread over a circular area of $1 cm^2$ and is dried in a small laboratory oven at 100 °C within 1 h. This oven should be used exclusively for that purpose.

The result of the carrier cleaning can be checked by TXRF itself. A spectrum of each carrier is recorded within 100 s and inspected for impurities. For example, Figure 4-3 shows element peaks arising from residual contaminations of about 10 pg. Such carriers should be sorted out and cleaned again in order to be suitable for ultratrace analysis. The success rate will be above 95% if a skillful operation is carried out.

4.1.2. Preparation of Samples

When analysis is to be performed of a small specimen representative of a larger amount of sample material, this material must first be homogenized. Furthermore, when a trace analysis is to be performed, the sample matrix should preferably be separated. In general, these operations are more difficult for solids than for liquids and more difficult for inorganic than for organic or biomaterials. A diagram of preparatory steps taken prior to TXRF analysis is given in Figure 4-4 [4].

Solid bulk materials can first be fragmented by cutting, sawing, crumbling, shredding, etc. The small pieces can then be ground down in a mill or mortar. The *fine powder* prepared in this way or received already pulverized is poured into a solution of water and ethanol. A fairly homogeneous suspension is

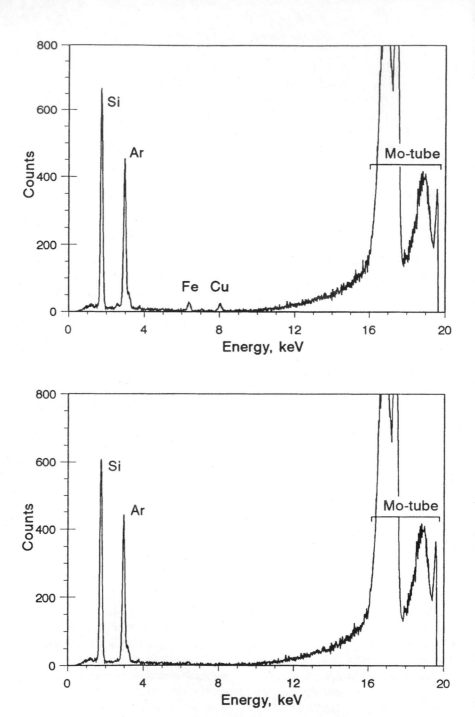

Figure 4-3. Spectra of a quartz glass carrier after cleaning. A first attempt (above) shows small impurities of Fe and Cu. The peak intensities of about 20 counts per 100 s come from an amount of only 10 pg. A second attempt (below) turns out to be successful.

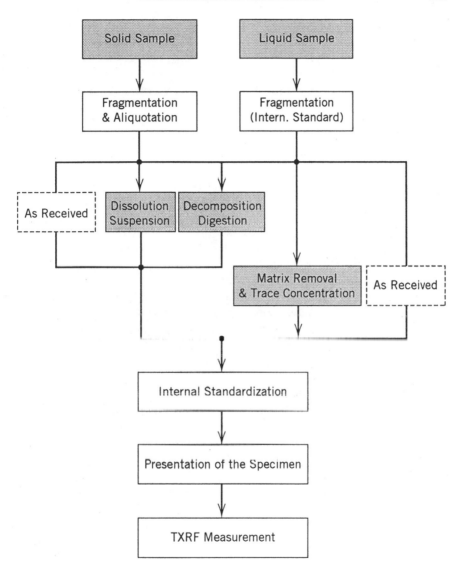

Figure 4-4. Diagram of preparatory steps taken prior to TXRF analysis. (After Prange [4].)

achieved by shaking or thoroughly mixing with a magnetic stirrer or by ultrasonication. Aliquots can be taken by a pipette and used as specimens for TXRF.

The solid material preferably in *pulverized* form can be dissolved either in total or in parts. Different inorganic and organic solvents are suited for that purpose but have to be ascertained for the individual material. After dissolution, an aliquot can be taken, pipetted on a carrier, and dried. In this way, the previous dilution of the sample is reversed and the detection of traces is not affected.

Biomaterials in particular can be subjected to various methods of sample preparation. Such methods are chosen to decompose the sample matrix and to transform it into a clear colorless solution. Samples can be ashed in an oxygen plasma driven by a radiofrequency discharge at a moderate temperature (possibly under high pressure). The ashed sample can be dissolved in nitric acid, for instance, and aliquots can be applied as specimens. A corresponding process can even be carried out on quartz glass carriers [5].

Furthermore, *mineral samples* and *biomaterials* can be digested in either an open or a closed vessel. The sample material is usually inserted into a Teflon or quartz glass vessel, and a strong acid or mixture of acids is added. Usually HNO_3 is chosen and also H_2O_2, HF, and HCl. Perchloric and sulfuric acid are not suited for TXRF since they do not evaporate afterward. Because of its shorter decomposition time, microwave heating is preferred to conventional heating. If the vessel is closed, the pressure will be increased and the decomposition accelerated. The temperature may come up to 300 °C and the pressure to 10^7 Pa. Digestion in a closed "bomb" is often called *pressure digestion*. It is particularly advantageous since element losses caused by volatilization are avoided. The decomposition is normally completed after 1 h or less, and the final solution can be used for analysis.

The methods just considered are intended for homogenization of solid bulks. They usually transform the solid materials into a liquid phase. The final solution contains the original sample in a certain dilution. For trace analyses, this solution should be concentrated, i.e., main compounds of the sample should be separated and the traces should be collected for analysis. Various methods already tested for other methods of elemental analysis, e.g., AAS and ICP-OES, have been utilized for this purpose. They are suitable for digested samples (e.g., metal digests) and for samples received as liquids.

First of all, a liquid can be concentrated by evaporation at elevated temperatures. Such drying is easily feasible with aqueous solutions. Other liquids are more or less suitable according to their boiling point and vapor pressure. A very gentle evaporation is made possible by freeze-drying. The sample is frozen just below the freezing point and simultaneously exposed to a fine vacuum of 0.1 Pa. When such an evaporation is performed for hours, it can raise the concentration of traces up to 10^6-fold.

Another simple method is based on volatilization of the matrix. By reduction of the volume and/or by addition of a liquid reagent, a chemical reaction is initiated whereby the matrix evaporates as a gaseous compound. The process usually runs at an elevated temperature.

The extraction of traces via different separating phases has also been successfully performed. Such a procedure is based on the addition of an inorganic acidic solution to the solution of the sample in an organic solvent. After thorough mixing, the two phases separate locally. The traces should thus have been transferred from the organic to the inorganic phase quantitatively while the sample matrix should have been left in the organic phase. The enriched inorganic phase is used for TXRF. In the case described here the traces are transferred from an organic to an inorganic phase, but this course can also be vice-versa.

A more generally applicable technique is the separation of traces by a chelating agent. After its addition, the metal traces are chemically bound as chelates. The solution is filtered by a special material whereby the metal complexes are adsorbed. Afterward, these complexes are eluted by an organic solvent and analyzed.

A summary of various methods of homogenization and preconcentration is given in Table 4-1 [6–25]. These have all been employed for TXRF analyses of different kinds of samples. All these methods of sample preparation must be carried out very carefully, and certain rules should be followed scrupulously. It is especially necessary to guard against *losses of* and *contamination by* elements; that can lead to serious systematic errors. Losses due to evaporation of volatile elements can be avoided by using closed vessels. Losses caused by adsorption on the vessel surface are reduced by use of quartz glass containers. Contaminations by reagents can be diminished by use of highly pure materials and only small quantities. Contaminations by vessels or instruments can be avoided by the exclusive use of quartz glass or Teflon and by careful cleaning. Contaminations from the ambient air can be prevented by working in a clean-bench. As a matter of principle, only one and the same vessel should be used for preparation, and pouring from one vessel into another should be avoided. For the decomposition of microquantities (< 10 mg), only small vessels with a volume of 1–3 mL should be used.

The addition of an internal standard is a further aid to eliminating certain systematic errors. With reference to a standard element, various nonspecific errors of preparation are compensated for, e.g., nonspecific losses. In order to compensate for a maximum of errors, the standard should be added to the sample as early as possible. This step, called *spiking* should be done after dissolution or suspension of the sample but before matrix separation or trace concentration if possible.

4.1.3. Presentation of a Specimen

Small amounts of a sample should be placed on a cleaned carrier and presented as a specimen. As already mentioned in Section 3.4.1, quartz glass is

Table 4-1. Different Methods of Sample Preparation Already Tested for TXRF Analyses of Various Kinds of Samples

Objective	Method	Kind of Sample	Physicochemical Process
Formation	Suspension	Aerosols, dust, ashes, sediments, pulverized biomaterials	Fine distribution of unsoluble particles in an aqueous solution (water + ethanol) by stirring or ultrasonication [6,7]
	Dissolution	Mineral oils, dust, air dust, filters	Solution in a solvent (water, chloroform, tetrahydrofurane [8-10] at moderate temperatures
Decomposition	Ashing	Air dust on filters, blood serum, mineral oils, digested fine roots	Ashing of the organic matrix (cellulose, protein) in an oxygen stream by heating (high-frequency heating) [11-14]
	Open digestion	Suspended matter, aerosols on filters, algae, blood, serum, tissue, ashes, mud, sediment	Solution of the sample in concentrated acids or mixtures (HNO_3, HF, HCl) by a chemical reaction at moderate temperature (infrared or microwave heating) [14-18]
	Pressure digestion	Dust on filters, lichen, minerals, soils, mussels, and fish	Like open digestion but in a closed vessel (PTFE bomb) at high temperatures (150–300 °C) and high pressures (10^6–10^7 Pa) [11,15,19]
Matrix removal	Drying	Rainwater, ultrapure acids	Evaporation of the matrix by warming up (possibly in a nitrogen stream [14,20,21]
	Freeze-drying	Drinking water, ultrapure water, ultrapure acids, organic solvents	Freezing of the matrix and evaporation under high vacuum [16,20]
	Volatilization	Digests of Si and SiO_2, sulfuric acid	Volatilization of Si as SiF_4 after addition of HF [22] Volatilization of S as SO_2 after addition of HI [22]
	Extraction	Dissolved high-purity iron, digested blood	Extraction of the Fe-matrix by MIBK [23] Extraction of the Fe-component by MIBK [17]
	Complexation	Rain- and seawater Digests of riverwater, blood and aluminum	Separation of the alkaline, alkaline earth, or aluminum components by complexation as carbamates, adsorption at Chromosorb or cellulose, and elution of the traces [15,19,20,24,25]

the most frequently used carrier for aqueous solutions and the most acidic and basic solutions. Although Plexiglas is a cheaper carrier, it is only suitable for aqueous solutions. After an electrostatically fixed protective foil is stripped off, the clean carriers can be used directly.

Only micrograms of solids ($1–200\mu g$) and microliters of liquids ($1–100\ \mu L$) yielding micrograms of dry residue can be used as specimens. The appropriate amount can easily be determined by means of the detector. The specimen should be restricted to an amount keeping the dead time below 50%. A more detailed consideration needed for quantitative analyses is given below in Section 4.3.3.

In most cases, the samples are received or prepared as liquids. These *liquids* can be pipetted onto the center of a circular carrier with a high degree of reproducibility. Pipettes with a volume between 5 and 50 μL are normally used (Eppendorf-Netheler-Hinz GmbH, Hamburg, Germany). The pipette tips, made of polypropylene, must not be touched and are used only once. The single *droplets* are dried by evaporation. For that purpose, the carriers are placed on a hot plate or under an infrared lamp. Droplets can also evaporate in a desiccator coupled with a diaphragm pump. But this technique is rather inconvenient. The hot plate method seems to be the most suitable. By this means, several droplets can be evaporated in rapid succession on top of each other. Organic solvents with a volume of $100–300\ \mu L$ can be dried up with the help of a small PTFE cylinder pressed against the carrier [24].

The residue may be seen as a bright spot of about 1–5 mm in diameter or may be hardly visible. Figure 4-5 demonstrates the simple technique of pipetting and additionally shows a small residue. This residue should be fairly

 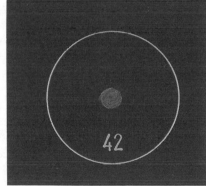

Figure 4-5. A droplet of 10 μL is pipetted on a carrier with a diameter of 3 cm (left). The droplet leaves a dry residue after evaporation (right).

dry, stable, and homogeneous and should stick to the carrier. For that reason, appropriate reagents are added either to the total sample or to the presented specimen. Multivalent alcohols and chelating agents are recommended [26, 27]. Nevertheless, a certain ring-shaped wall may form that can affect the quantitative analysis. This disruptive effect, however, can be compensated by the addition of an internal standard.

Besides real liquids, *fine powders* or pulverized solid materials can be presented via a droplet. They must first be prepared as a suspension that should be stirred thoroughly. After that, droplets of about 10 μL can be presented as specimens. But fine powders can also be applied directly. A spatula made of PTFE can be used to take a small amount of the powder and dust it on a carrier. The small grains usually stick to the carrier but can be removed by knocking the carrier when too large an amount has been taken.

Furthermore, *pigments* of oil paints can be sampled by means of a cotton bud, often called a Q-tip. Some material is rubbed off the painted surface, and the loaded Q-tip is dabbed onto a carrier [28]. Microgram amounts can be transferred in this way. Of course, this technique is also applicable to other mineral or oxide powders.

Air dust or more generally *aerosols* can be collected on filters, which have to be ashed or digested prior to analysis. But they can also be collected directly on the carriers recommended for TXRF. These carriers can be used as impaction plates in a device called an *impactor*. The dust particles taken by an airstream are deposited according to their inertia. The loaded carriers can be directly used for analysis [10, 29].

Individual particles like crumbs, grains, or fibers of a few micrograms can simply be put right onto the carrier. A wooden toothpick should be used for the manipulation. Such samples are especially suitable for a qualitative analysis by TXRF.

Small amounts of *solid bulk* samples can be removed by laser ablation [30]. The laser beam is focused by an objective with a long focal length. Material is molten, even evaporated, and is thrown out from a small crater about 10 μm in diameter. The emitted material is deposited on a quartz glass carrier that should be positioned between the objective and the sample, just above the latter, and that should be penetrated centrally by the laser beam. The material deposited on the carrier can be used as a specimen for TXRF. This sampling technique is capable of providing a microdistribution analysis, e.g., by a line scan. It can also be used for a depth-profiling analysis. For that purpose, the sample should first be abraded at an angle of < 1°.

When only a survey but not a local analysis is required, another simple technique can be applied: *solid samples* can be rubbed on a hard quartz glass carrier in a single stroke; or, the other way round, a quartz glass carrier can be rubbed on a fixed object that has a lower degree of hardness. In either case, a small amount of sample material will be smeared onto the carrier and can be used as a specimen [7]. This technique is especially useful for large finished products that are hardly accessible by other means.

Organic materials and *biomaterials* are normally decomposed by ashing, combustion, and/or wet chemical digestion. But such materials can also be freeze-cut by a microtome. The frozen sections are placed on a carrier, dried, and directly analyzed. For quantification, an internal standard can be added to the section afterward [31].

The liquid and solid samples enumerated so far are presented for micro- or trace analysis. In addition, there are samples suitable for surface or thin-layer analysis by TXRF. *Wafers* not yet patterned can directly be applied as flat disks provided that an appropriate sample holder is available. Uncoated wafers can be subjected to a surface analysis; coated wafers, to a thin-layer analysis. One must be aware that the actual specimen is defined by the detector's range of vision, i.e., by the projection of its front area onto the wafer disk. Only this region is analyzed. In order to examine the total wafer, a displacement device is necessary. Such a thin-layer analysis is also possible when *thin films or layers* are deposited on a flat substrate, e.g., quartz glass. The samples can directly be used as specimens for TXRF. This analysis can be of use for monitoring a sputter or vapor-phase decomposition process, for example.

4.2. RECORDING AND INTERPRETATION OF SPECTRA

The actual measurement consists in recording the spectra. The carriers prepared for that purpose are put into sample holders. These plastic slides are either pushed directly into the fixed measuring position or first loaded in a sample changer and afterwards pushed into that position automatically one after another. The X-ray tube, filter, first reflector (monochromator), and carrier must be set in a certain combination and geometry as recommended for analysis. The tube voltage, tube current, and preset live time have to be selected appropriately, in accord with definite rules. After these preparations, the spectra are recorded and stored. The subsequent interpretation aiming at a qualitative analysis is based on the detection of individual peaks, identification of these peaks, and finally determination of elements inferable from these peaks. A particularly useful mode of qualitative analysis is called *fingerprint analysis*, which discerns the similarity of two complete spectra by superposition.

4.2.1. The Instrumental Setup

X-ray tubes, foil filters, and reflectors (monochromators) are components that can usually be chosen or exchanged for the particular operation. Of the several

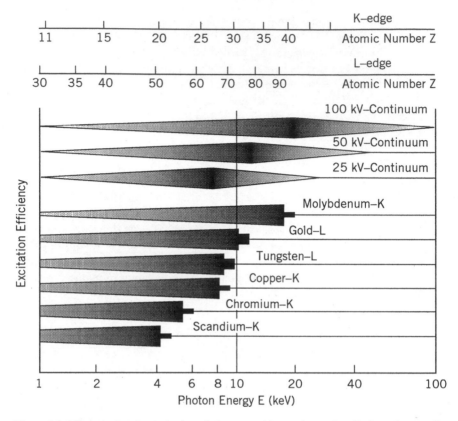

Figure 4-6. Efficiency for the excitation of elements with atomic number Z, dependent on the photon energy E. The excitation can be performed by the continuum (100, 50, 25 kV) and by the K- or L-peaks of different X-ray tubes (Mo, Au, W, Cu, Cr, Sc). The upper two scales show the energetic position of the K- and L-absorption edges of the analytes.

combinations that have already been tested, some are commonly used whereas others are appropriate for specific applications. The aim is an effective excitation of the sample. But the different elements require different conditions for an optimum excitation. Consequently, a compromise solution is generally necessary for multielement analyses.

Excitation is generally performed by line-focus X-ray tubes. The continuous radiation as well as the characteristic radiation of the anode material can contribute to excitation, but with distinct differences in efficiency. This is illustrated in Figure 4-6. Excitation is only possible by photons with an energy that exceeds the respective absorption edge of the element. The continuum is effective for excitation within a wide range of photon energies. It can generate the K- and/or L-peaks of nearly all elements with atomic number Z if the operating voltage is above 25 kV. Maximum efficiency will be achieved if the edge energy is about a fifth or a third of the energy that corresponds to the applied voltage.

The characteristic radiation, on the other hand, will be highly effective if the peak energy of the tube anode just equals the absorption edge of an analyte. Excitation above this value is not possible; excitation below this value is steadily decreasing. As can be seen in Figure 4-6, a Mo-tube operated above 50 keV is suitable for the excitation of all elements. But a second tube is desirable to excite the lighter elements ($Z < 20$), with their K-peaks, and the medium heavy elements ($40 < Z < 60$), with their L-peaks, more effectively. The tubes may be exchangeable, but it is more convenient when both are permanently installed.

The tube spectrum is usually processed by a metal foil acting as a specific absorption filter, by a totally reflecting mirror acting as a low-pass filter, or even by a natural crystal or multilayer acting as a Bragg monochromator (see Section 3.3). These components are used to further reduce the background of the spectra or to avoid a blurring of angle-dependent intensity profiles. For micro- and trace analyses, the spectral background has to be lowered, which can be achieved by a combination of filters. For surface and thin-layer analyses blurring must be suppressed, which is preferably done by a natural crystal or multilayer. Metal foils clamped in holders can easily be positioned and replaced, but mirrors or Bragg reflectors have to be carefully adjusted and should be left in this position continuously.

For micro- and trace analyses, three different excitation modes might be recommended:

a. A W-tube operated at 50 kV, a Ni-foil of 100 μm, and a quartz glass mirror set at a glancing angle of 0.05° (pass energy 35 keV)

b. A Mo-tube operated at 50 kV, a Mo-foil of 50 μm, and a quartz glass mirror set at 0.09° (pass energy 20 keV)

c. A W-tube operated at 25 kV and a Cu-foil of 10 μm; with respect to mode a, only the foil has to be exchanged and the voltage reduced (the tube and the mirror can be kept unchanged)

The three modes can easily be interchanged when the two required tubes are both installed. Three further but similar modes of excitation were tested by Knoth et al. [32]. They only need a single W-tube and a combination of two multilayers. But this arrangement has to be adjusted back and forth in order to change the mode.

The primary spectrum processed in the three modes a–c is sketched in Figure 4-7. It illustrates that excitation is mainly realized by a broad spectral

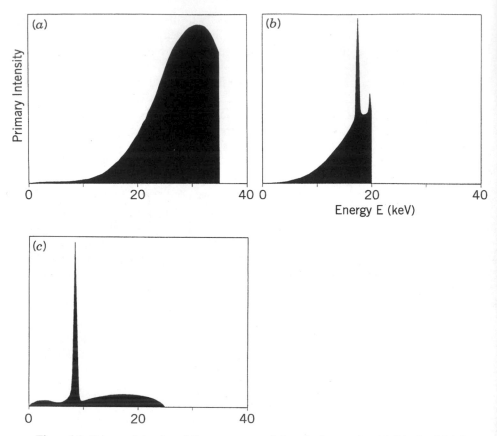

Figure 4-7. Primary intensity of three recommended excitation modes: (a) W-tube, Ni-foil, low-pass filter; (b) Mo-tube, Mo-foil, low-pass filter; (c) W-tube, Cu-foil.

band (a), by the Mo-$K\alpha$ and -$K\beta$ peak (b), and by the W-$L\alpha$ peak (c). Mode a can be used for excitation preferably in the energy range between 9 and 30 keV; mode b is preferable for excitation between 4 and 16 keV; and mode c is suitable for excitation energies between 2 and 8 keV. Figure 4-8 demonstrates the efficiency of the three modes for the detection of elements with atomic number Z; they are detected either by their K-peaks or their L-peaks.

For surface and thin-layer analyses, only a single peak or a small band of the continuum is cut out of the spectrum by means of a real monochromator. Natural crystals or multilayers tried and tested for this purpose are listed in Table 3-1. If two X-ray tubes, e.g., a Mo- and a W-tube, are disposable, the two peaks Mo-$K\alpha$ and W-$L\alpha$ may be used again. But the W-$L\beta$ peak is normally

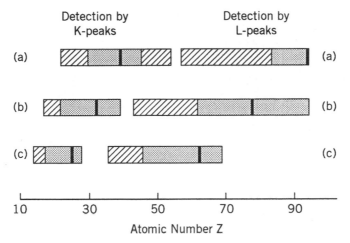

Figure 4-8. Efficiency of three special modes of excitation applied for the detection of elements with atomic number Z. Lower-Z elements are detected by their K-peaks; higher-Z elements, by their L-peaks. The three modes a, b, and c have already been described: *hatching*, moderate efficiency; *dotting*, high efficiency; *solid black*, optimum efficiency.

preferred to the W-$L\alpha$ peak because it has the advantage of additionally exciting the elements Cu and Zn. If only one X-ray tube is available, two separate monochromators may be useful, installed in the same instrument [33]. In this way, lighter and heavier elements can be detected with a comparable efficiency.

4.2.2. Recording the Spectrograms

Once chosen, the instrumental setup is usually kept unchanged. But some devices have to be put in operation over and over again, and the operational parameters may be varied according to the sample and the analytical problem.

The operating voltage of the X-ray tube will usually be set to a *fixed* value if a quantitative analysis is also planned. The sensitivity values needed for quantification are not applicable to any but a fixed operating voltage. As already mentioned, values of 50 and 25 kV are commonly used settings. The operating current of the tube, on the other hand, can be selected in a wide range, normally between 5 and 60 mA. Its particular setting is irrelevant even for quantification. But an upper limit is determined by the dead time of the detection unit. The current has to be limited to such a value that the dead-time losses are kept below 50%. For trace analyses, this current should be chosen if at all possible. If the applied current is at first set too high, it should be lowered; otherwise the sample mass must be reduced. Finally, the count-rate meter should indicate a value between 300 and 12,000 cps.

Only a few parameters have to be set for the EDS. The energy scale can normally be restricted to 10, 20, and 40 keV, respectively. The decision depends on the analytical problem and its approach. Excitation mode a requires a 40 keV recording; mode b, the usual 20 keV setting; and mode c, the 10 keV setting. If the MCA has the usual number of 1000 channels, the channel width will consequently be fixed at 40, 20, and 10 eV, respectively. A shaping time of 4 μs is usually selected. It should be increased when a better spectral resolution is wanted and should be decreased when a higher count rate is desirable. Finally, the energy axis has to be calibrated or recalibrated. This procedure should be repeated twice a day. For that purpose, some special carriers are prepared and always reused. They are loaded with the residue of an aqueous solution of certain standard elements. Molybdenum is prefered for excitation mode a, iron for mode b, and titanium for mode c.

After a warm-up phase of about 1 h, the individual spectra of the different samples can be recorded. A live time is preset between 10 and 1000 s; a 100 s setting is commonly selected, and a 1000 s setting is only chosen for the detection of ultratraces. The increasing spectra can be observed on a color display during measurement. At the end, the spectra are coded and stored for subsequent evaluation and analysis. Of course, all samples are first examined in a certain excitation mode. After such examination, the mode is changed and the samples are examined in the next mode.

4.2.3. Qualitative Analysis

The capability of performing a simultaneous multielement analysis is the most obvious virtue of an EDS, highly useful when one needs to get an idea of the composition of a sample that is initially completely unknown. It enables a further analytical strategy without overlooking any essential element. The necessary qualitative analysis is based on the interpretation of the spectra. Each spectrum is treated and evaluated as a whole. A data processing system is generally used to provide a rapid and extremely convenient evaluation. Hardware and software already used for conventional EDS are similarly utilized here. The spectra can be spread or compressed on both the energy axis and the intensity axis, and can be rolled back and forth along the energy axis. They can also be smoothed and corrected for escape peaks. A second spectrum can be adjusted to the intensity of the first spectrum and stripped in order to cancel out certain overlappings. Markers can be recalled in order to indicate the position of the peaks of any selected—or expected—element.

4.2.3.1.   *Element Detection*

In general, X-ray spectra are quite simple and show only a few peaks, much fewer than ultraviolet (UV) spectra do. There are only about 500 distinct peaks for about 80 elements that can be detected by EDS (atomic number > 11). All these peaks are in accord with the Moseley law with regard to their position, and peaks of the K-, L-, or M-series keep to certain intensity ratios (see Section 1.2.1). Principal peaks with a high intensity are listed in Table 4-2. More extensive tables that also include weak peaks are available in the literature [34, 35].

A qualitative analysis is intended to detect elements by means of their characteristic peaks in the spectrum of a specimen. This problem is generally

Table 4-2. Principal Peaks of X-ray Spectra Obtained from Elements with Atomic Number Z^a

Z	Elem.	$K\alpha_1$	$K\beta_1$	$L\alpha_1$	$L\beta_1$	$L\beta_2$	$L\gamma_1$	$M\alpha$	$M\beta$
6	C	0.277							
7	N	0.392							
8	O	0.525							
9	F	0.677							
10	Ne	0.848							
11	Na	1.041	1.067						
12	Mg	1.253	1.295						
13	Al	1.486	1.553						
14	Si	1.740	1.829						
15	P	2.013	2.136						
16	S	2.307	2.464						
17	Cl	2.622	2.815						
18	Ar	2.957	3.190						
19	K	3.313	3.589						
20	Ca	3.691	4.012	0.341	0.345				
21	Sc	4.090	4.460	0.395	0.400				
22	Ti	4.510	4.931	0.452	0.458				
23	V	4.951	5.426	0.511	0.519				
24	Cr	5.414	5.946	0.573	0.583				
25	Mn	5.898	6.489	0.637	0.649				
26	Fe	6.403	7.057	0.705	0.718				
27	Co	6.929	7.648	0.776	0.791				
28	Ni	7.477	8.263	0.851	0.869				
29	Cu	8.046	8.904	0.930	0.950				

(*Contd.*)

Table 4-2. (*Contd.*)

Z	Elem.	$K\alpha_1$	$K\beta_1$	$L\alpha_1$	$L\beta_1$	$L\beta_2$	$L\gamma_1$	$M\alpha$	$M\beta$
30	Zn	8.637	9.570	1.012	1.034				
31	Ga	9.250	10.263	1.098	1.125				
32	Ge	9.885	10.980	1.188	1.218				
33	As	10.542	11.724	1.282	1.317				
34	Se	11.220	12.494	1.379	1.419				
35	Br	11.922	13.289	1.480	1.526				
36	Kr	12.648	14.110	1.586	1.636				
37	Rb	13.393	14.959	1.694	1.752				
38	Sr	14.163	15.833	1.806	1.871				
39	Y	14.956	16.735	1.922	1.995				
40	Zr	15.772	17.665	2.042	2.124	2.219	2.302		
41	Nb	16.612	18.619	2.166	2.257	2.367	2.461		
42	Mo	17.476	19.605	2.293	2.394	2.518	2.623		
43	Tc	18.364	20.615	2.424	2.536	2.674	2.792		
44	Ru	19.276	21.653	2.558	2.683	2.835	2.964		
45	Rh	20.213	22.720	2.696	2.834	3.001	3.143		
46	Pd	21.174	23.815	2.838	2.990	3.171	3.328		
47	Ag	22.159	24.938	2.984	3.150	3.347	3.519		
48	Cd	23.170	26.091	3.133	3.316	3.528	3.716		
49	In	24.206	27.271	3.286	3.487	3.713	3.920		
50	Sn	25.267	28.481	3.443	3.662	3.904	4.130		
51	Sb	26.355	29.721	3.604	3.843	4.100	4.347		
52	Te	27.468	30.990	3.769	4.029	4.301	4.570		
53	I	28.607	32.289	3.937	4.220	4.507	4.800		
54	Xe	29.774	33.619	4.109	4.422	4.720	5.036		
55	Cs	30.968	34.981	4.286	4.619	4.935	5.279		
56	Ba	32.188	36.372	4.465	4.827	5.156	5.530		
57	La	33.436	37.795	4.650	5.041	5.383	5.788	0.838	0.854
58	Ce	34.714	39.251	4.839	5.261	5.612	6.051	0.883	0.902
59	Pr	36.020	40.741	5.033	5.488	5.849	6.321	0.929	0.949
60	Nd	37.355	42.264	5.229	5.721	6.088	6.601	0.978	0.996
61	Pm	38.718	43.818	5.432	5.960	6.338	6.891		
62	Sm	40.111	45.405	5.635	6.204	6.586	7.177	1.081	1.100
63	Eu	41.535	47.030	5.845	6.455	6.842	7.479	1.131	1.153
64	Gd	42.989	48.688	6.056	6.712	7.102	7.784	1.185	1.209
65	Tb	44.474	50.374	6.272	6.977	7.365	8.100	1.240	1.266
66	Dy	45.991	52.110	6.494	7.246	7.634	8.417	1.293	1.325
67	Ho	47.539	53.868	6.719	7.524	7.910	8.746	1.347	1.383
68	Er	49.119	55.672	6.947	7.809	8.188	9.087	1.405	1.443
69	Tm	50.733	57.506	7.179	8.100	8.467	9.424	1.462	1.503

(*Contd.*)

Table 4-2. (*Contd.*)

Z	Elem.	$K\alpha_1$	$K\beta_1$	$L\alpha_1$	$L\beta_1$	$L\beta_2$	$L\gamma_1$	$M\alpha$	$M\beta$
70	Yb	52.380	59.356	7.414	8.400	8.757	9.778	1.521	1.567
71	Lu			7.654	8.708	9.047	10.142	1.581	1.631
72	Hf			7.898	9.021	9.346	10.514	1.644	1.697
73	Ta			8.145	9.342	9.650	10.893	1.709	1.765
74	W			8.396	9.671	9.960	11.284	1.775	1.835
75	Re			8.651	10.008	10.274	11.683	1.842	1.906
76	Os			8.910	10.354	10.597	12.093	1.914	1.978
77	Ir			9.174	10.706	10.919	12.510	1.980	2.053
78	Pt			9.441	11.069	11.249	12.940	2.050	2.127
79	Au			9.712	11.440	11.583	13.379	2.123	2.204
80	Hg			9.987	11.821	11.922	13.828	2.195	2.282
81	Tl			10.267	12.211	12.270	14.289	2.270	2.362
82	Pb			10.550	12.612	12.621	14.762	2.345	2.442
83	Bi			10.837	13.021	12.978	15.245	2.422	2.525
84	Po			11.129	13.445	13.338	15.741	2.502	2.618
85	At			11.425	13.874	13.705	16.249	2.582	2.707
86	Rn			11.725	14.313	14.077	16.768	2.663	2.795
87	Fr			12.029	14.768	14.448	17.300	2.745	2.882
88	Ra			12.338	15.233	14.839	17.845	2.826	2.968
89	Ac			12.650	15.710	15.227	18.405	2.909	3.054
90	Th			12.967	16.199	15.621	18.979	2.996	3.145
91	Pa			13.288	16.699	16.022	19.565	3.082	3.239
92	U			13.612	17.217	16.425	20.164	3.170	3.336
Intensity ratios		100 : 15		100 : 50 : 20 : 10				100 : 50	

Source: After Johnson and White [35].

[a] The photon energy is given in keV ($E < 60$ keV). The peaks appear with an approximate intensity ratio given in the last line of the table.

solved in three steps. First, the peaks are detected by their appearance and their centroid positions are localized (e.g., peaks at 8.05 and 8.91 keV). Secondly, these peaks are identified as X-ray peaks by comparison with tables or data files (for the given example: Cu-$K\alpha$ or Ir-Ll and Cu-$K\beta$ or Os-$L\alpha$, respectively). Thirdly, from these data the presence of particular elements is deduced (for the given example: Cu). These three steps—peak localization, peak identification, and element deduction—can also be carried out automatically by a search program. But the results should always be checked by an experienced operator.

The first step—*peak localization* is made complicated by the poor resolution of an energy-dispersive detector and by the statistical fluctuations of

X-ray spectra. The position of the peaks cannot therefore be determined exactly. Strong peaks are localized with an uncertainty of $\pm$ 10 eV in energy; weak peaks, with as much as $\pm$ 50 eV. Weak peaks may be overlapped by strong peaks and may even be overlooked. Overlappings, or "interferences," frequently occur in the energy range below 10 keV. But weak peaks of trace elements may also disappear in the fluctuations of the spectral background. This is a problem having to do with detection limits and is further treated in Section 5.1.2.

The second step—*peak identification*—is hampered in consequence of the aforementioned uncertainty. Frequently, not only one tabulated peak fits an energy position, but two or even more peaks may be assigned to the permissible bandwidth. In such cases, too many peaks may be assigned—except for those cases where two peaks really exist with a strong interference. For example, the $K\alpha$-peaks of medium-Z elements overlap with the $K\beta$-peaks of the neighboring elements in the periodic table. Close interferences are also produced by S-$K\alpha$, Mo-$L\alpha$ ($\Delta E = 15$ eV), and Pb-$M\alpha$ ($\Delta E = 38$ eV), by Ti-$K\alpha$ and Ba-$L\alpha$ ($\Delta E = 43$ eV), or by As-$K\alpha$ and Pb-$L\alpha$ ($\Delta E = 8$ eV). More seldom, not even one peak may be found with the help of any table in use. These generally weak peaks may be artifacts of the EDS, e.g., sum peaks or escape peaks (see Section 3.5). Moreover, they can represent "satellite" and "forbidden" peaks of main constituents, which may be missing in some tables (see Section 1.2.1) Satellite peaks arise from doubly ionized instead of simply ionized atoms and can possibly be observed as two small peaks near a strong $K\alpha_1$, $K\alpha_2$ doublet (e.g., $K\alpha_3$, $K\alpha_4$ or $K\alpha_5$, $K\alpha_6$). Forbidden peaks will arise from outer energy levels if these are close together and the selection rules for quantum transitions are no longer stringent. Forbidden peaks can be observed as small peaks near strong $K\beta$- or $L\beta$-peaks (e.g., $K\beta_4$ or $L\beta_8$).

The last step—*element deduction*—is actually rather easy. The first necessary condition for the detection of an element is the appearance of its strongest principal peak ($K\alpha$ or $L\alpha$). But an element will only be considered to be unambiguously detected if two of its strong principal peaks are identified ($K\alpha$ and $K\beta$, or $L\alpha$ and $L\beta$, or $K\alpha$ and $L\alpha$) and are in a reasonable intensity proportion. This rule can be followed for main and additional constituents of the specimen but not for traces. In trace analysis, only the strongest peak ($K\alpha$ or $L\alpha$) should be called for. Attention has to be paid to satellite and forbidden peaks of main constituents. They must not be confused with a weak peak of a trace element.

It should further be taken into account that some strong peaks can also come from other sources and not only from the specimen itself. The Si peaks can arise from a quartz glass or silicon carrier, the Ar peaks are normally caused by ambient air, and the peaks of Mo and W can be due to the X-ray tube. Under such circumstances, these peaks must not be used for the detection of elements of the actual specimen. One must be aware of all these phenomena, but the difficulties should not be overestimated. A careful operator with some experience will find no undue problem and will make no grave mistakes.

4.2.3.2. *Fingerprint Analysis*

A first approximation to a *quantitative* analysis is achieved by a fingerprint or signature analysis. Two complete spectra are compared for that purpose and are checked for equality *in toto* or at least in part. Energy-dispersive spectra are highly suitable for this type of analysis since the complete spectra are always recorded. But both spectra to be compared must be recorded in the same excitation mode with the same energy range, and must be stored in the computer memory.

Fingerprint analysis is carried out in a "compare" mode when the two spectra are jointly displayed on the video screen. The first spectrum may be displayed in bars; the second, in dots. With regard to the intensity, both spectra can be adapted to each other by spreading the weaker spectrum or compressing the stronger one *in toto*. After that, the two spectra are compared like two patterns and equality or similarity is accepted or rejected. This simple yes/no decision can be relevant for certain parts of the spectrum or even for the total spectrum.

In a most simple case of fingerprint analysis, the presence of a certain element may be ascertained by overlapping the spectra of the unknown sample and of the pure element. For that purpose, a spectral library of individual pure elements is useful. It can be set up by aqueous solutions, single grains, and metal filings of the pure elements or even by pure gases pouring into the sample chamber. If a complete library of (nearly) all the elements is available, this technique can provide a qualitative analysis. To this end, the presence of all these elements in the specimen has to be checked one after the other.

In most cases, however, the presence of only a few elements is confirmed and even their approximate concentrations are estimated. Usually, the spectra of the specimen in question and of a specimen defined as the reference are compared. Equality of the spectra indicates the identity of both specimens; dissimilarity means just the opposite. Such fingerprint analyses are very useful for some typical applications. They are used to prevent a mixup of alloys before processing, to distinguish between genuine art objects and fakes, and to scrutinize suspect materials in forensics.

4.3. QUANTITATIVE MICRO- AND TRACE ANALYSES

Ideally a linear relationship exists between the amount or mass m_x of an analyte element x excited and the X-ray intensity N_x recorded. If a constant volume of solutions is applied for analysis, this relationship can be written as

$$N_x = B_x c_x \tag{4-1}$$

where N_x is the background-corrected net intensity of the principal peak or peaks of the analyte x; B_x, a proportionality factor called absolute sensitivity; and c_x, the volume concentration of the analyte. Instead of the fluorescence intensity I of Chapter 2, emitted from a sample, this net intensity is indicated by N, as it means total counts recorded in a preset live time of the detector.

The plot of equation (4-1) gives a straight line called the *calibration straight line*. Its slope is determined by the sensitivity B_x. Individual elements differ by their particular sensitivity, or the slope of their straight line. The excitation and detection have to be definite and kept constant, and the chemical composition and the physical state of the sample matrix should not affect the calibration.

In XRF analysis, this ideal case is fairly well approximated for, e.g., thin layers or thin films, while notorious deviations occur for thick layers or bulk samples. Such deviations are called *matrix effects*. In TXRF, only small residues of liquids, small amounts of powders, thin sections or layers, and individual particles are subject to analysis. Such samples with a tiny mass and thickness also meet the conditions for the ideal case. Quantification can then be done by simply adding a standard to the sample and by subsequently using it as an *internal* standard. When only one element has to be determined, this element itself can be used as the standard element. This method is well known as the standard-addition method from atomic absorption spectrometry (AAS). When, however, several elements must be determined, an additional element is added that was previously not present in the sample. This method is appropriate for a multielement determination and is normally used for TXRF. It is based on the sensitivity values of the different elements and already known from other methods of atomic spectroscopy.

Both variants of quantification first require the determination of the net intensity of principal peaks in the X-ray spectrum. They may arise from one or more analyte elements and from the internal standard element. Besides, the sensitivity values of these elements should be known from preparatory measurements. But this work has to be carried out only once as a kind of calibration. Each subsequent quantification is generally easy and reliable, and consequently TXRF has great advantages over conventional XRF.

Of course, there are some conditions for the applicability of internal standardization. But the limitations can easily be observed. They apply to the sample amount used for micro- and trace analyses. On the other hand, a real surface or thin-layer analysis should be carried out in accord with its own regulations, which are treated in Section 4.4.

4.3.1. Prerequisites for Quantification

The net intensities of the principal peaks have to be determined in each spectrum. This can be done following the common practice in EDS. The sensitivity values for these peaks or the respective elements, however, need to be determined quite seldom. This is done for any particular excitation mode and is only repeated after a repair or replacement of instrumental components.

4.3.1.1. Determination of Net Intensities

An X-ray spectrum generally shows several peaks on a structured background, some of which may partly overlap. These peaks have already been identified and assigned to certain elements by means of a previous qualitative analysis. Each element x is consequently represented by two or three principal peaks, or at least by one. The net intensity N_x of the strongest peak or possibly of some additional peaks must be determined now for each analyte element x. This intensity is defined by the area below the peaks but should be corrected for the spectral background and for the overlapping of a neighboring peak. It may be determined by summation of all the counts being accumulated in the individual channels representing the peaks and by a subsequent subtraction of the background and a possibly overlapping peak. This can be done semiautomatically after settling of "windows" that limit each peak to a lower and upper border of the photon energy. The correction is made by subtraction of a trapezoidally shaped area below the peak and between the borders. This method is a rough approximation that is only suitable for strong peaks on a low background and without strong overlapping. In this case, the window width can be chosen to be two or three times the peak width (FWHM) so that the correction is appropriate. In most cases, however, a severe error will result from use of this method, which is unacceptable.

For that reason, other methods have been developed that can also be applied automatically. The first corrects for the background and is based on a Fourier transformation of the spectrum. This mathematical procedure results in three transformed but significantly different parts of the spectrum. The low-frequency part represents the background, the medium-frequency portion corresponds to the peaks, and the high-frequency part is equivalent to the noise of the background. The low- and high-frequency parts are removed by a mathematical filter and only the medium-frequency part is retransformed. This method, also called filter technique, gives a background-corrected and smoothed spectrum only containing the spectral peaks. The net intensity can easily be determined now, but overlaps still exist and are troublesome.

For the latter reason, a second method was developed which can be applied to background-free spectra in order to correct for the overlaps. It is based on the principal peaks of pure elements and fits them to each spectrum in question. Consequently, it needs a spectral library like that required for fingerprint analyses in Section 4.2.3.2. Each spectrum of those pure elements detected by a previous qualitative analysis is reduced in peak height until its principal peaks approximately fit with the corresponding peaks of the spectrum in question. After that, the single spectra are added and a last adaptation to the relevant spectrum is made by a least square fit. Finally, the intensity values of the adapted spectra of the pure elements can be read. Each intensity value will preferably include all principal peaks of the respective element, e.g., the K-peaks and/or the L-peaks. The final result is free from the spectral background and from any overlapping, and also escape peaks are taken into account. Apparently, both methods of correction successively applied give highly reliable values of net intensities.

4.3.1.2.   Determination of Relative Sensitivities

The linear relationship (4-1) between the XRF intensity of an analyte element and its concentration will be valid if the analyte is part of a small specimen deposited on a glass carrier. The plot of the measured net intensity vs. the concentration of this element gives a calibration straight line. Its slope is called *absolute sensitivity*. As noted earlier, different elements generally have different slopes or sensitivities. The ratios of these absolute sensitivities with reference to a specific element are called *relative sensitivities* and can be determined by calibration. The mode of excitation chosen for that purpose has to be precisely defined. The X-ray tube, applied voltage, filter, and/or monochromator as well as the geometry of the equipment in use must be fixed exactly. Any altered mode of excitation requires a new calibration and leads to a new set of relative sensitivities.

For the determination of relative sensitivities, different standard solutions are used that are commonly employed for calibration in AAS or ICP spectral analysis. They are obtainable at various suppliers (e.g., E. Merck, Darmstadt, Germany; Aldrich-Chemie GmbH, Steinheim, Germany; Johnson Matthey GmbH—Alfa Products; Karlsruhe, Germany). For TXRF, either complete multielement standards are chosen or single-element standards are mixed together. The stock solutions should be diluted by an aqueous solution (5% nitric acid, suprapure) to a concentration level of 1–10 μg/mL for the individual elements. The finally applied standard solutions should have a volume of about 1 mL and should contain two to six different elements. One element, e.g., Co or Cu, which is chosen as the reference element, must be present in all standard solutions.

A droplet of each solution with a volume of about 2–10 μL is pipetted onto a cleaned carrier and dried by evaporation. From the residues, spectra are recorded in one or several definite excitation modes, as demonstrated in Figure 4-9a–c. The net intensity for the individual elements is determined and the relative sensitivities are calculated by the formula

$$S_j = \frac{N_j/c_j}{N_{rf}/c_{rf}} S_{rf} \tag{4-2}$$

where S is the relative sensitivity; N, the net intensity, and c, the concentration—of the different elements j or the reference element rf, as indicated. The quantity S_{rf} can generally be set to 1 since only *relative* sensitivities have to be determined. Actually, a single determination per element is sufficient. But it is preferable to apply several solutions with different concentrations and to repeat the measurements. The averaged values of each element should have a coefficient of variation of $< 5\%$.

Figure 4-10 provides an overview of relative sensitivities of detectable elements, determined for the three excitation modes defined in Section 4.2.1. The semilogarithmic plot describes their dependence on the atomic number of the elements, detected by either their K- or their L-peaks. The six curves span 3 orders of magnitude with a steady course. Each curve shows a maximum sensitivity reached for a particular element in accordance with Figure 4-8.

Above all, it should be emphasized that the relative sensitivities are independent of the matrix of the applied samples. For the case described above, the matrix water was separated by evaporation and only oxides, hydroxides, and further light compounds of the analyte elements remained as residues. If the original solution, however, is a saline or a gelatinous solution, a mineral or an organic matrix, respectively, will form the residue. Nevertheless, the relative sensitivities of elements determined with these matrices correspond to each other with a deviation of $< 8\%$ [36].

Relative sensitivities can even be calculated from theory. In accord with equation (2-37), the sensitivity can be expressed by

$$S_j = K g_j \omega_j f_j (\tau/\rho)_{j, E_0} \tag{4-3}$$

where S_j is the sensitivity of the analyte j; g, the relative emission rate of the respective element peak in its series, ω, is the fluorescence yield of the selected K- or L-peaks; f, the jump factor of the respective absorption edge; and $(\tau/\rho)_{E_0}$ the photoelectric absorption coefficient for the primary beam with a photon energy E_0. The constant K can be determined with the help of a reference element for which S_{rf} is set to be 1. The fundamental data $[g, \omega, f,$ and $(\tau/\rho)]$ can be taken from tables. The efficiency of the detector and the transmission of the air path between the sample and the detector are assumed here to be equal for the individual elements, but they can easily be determined and taken into

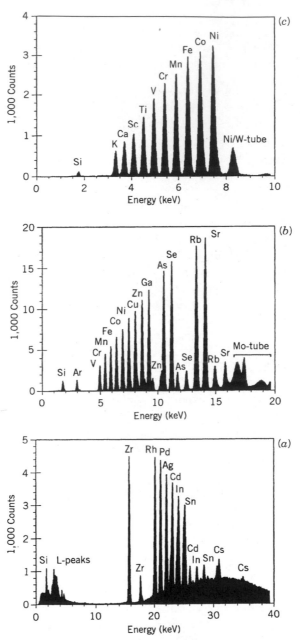

Figure 4-9. Multielement spectra obtained for the three special modes of excitation described in Section 4.2.1. The peaks of all standard elements arise from an applied amount of 1 ng per element. The different peak heights reflect the different sensitivities. Excitation is mainly performed by (a) the W-continuum at 30 keV; (b) the Mo-K peaks; and (c) the W-$L\alpha$ peak.

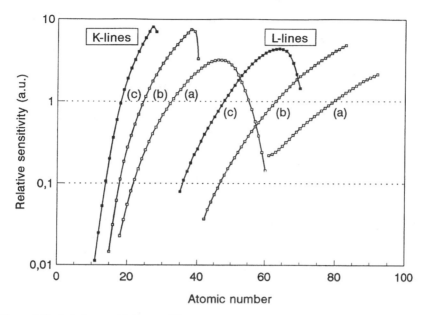

Figure 4-10. Relative sensitivities on different elements in dependence on their atomic number Z. Three excitation modes—a, b, and c—were applied as described in Section 4.2.1. The elements were detected either by their K-peaks (medium Z-values) or by their L-peaks (high Z-values).

consideration. The results correspond to the values experimentally determined within a margin of 9%, evidence for the validity of the theory and the absence of matrix effects [36].

It can be concluded that the sensitivity values of those elements for which standards are not available can be obtained by calculation or at least by interpolation between known values. Relative sensitivities can even be transferred from one instrument to another if the equipment is the same.

4.3.2. Quantification by Internal Standardization

In TXRF, quantification is generally carried out by internal standardization. This easy and reliable method can be employed since small or minute specimens are used for analysis. Two different methods can be chosen to this end: the first is suitable for a quantitative determination of a single element; the second is generally recommended for a multielement determination, which is the actual task of TXRF.

4.3.2.1.   Standard Addition for a Single Element

The first method is known as *standard addition* and is generally used for single element determinations in AAS [37]. A few aliquots (three to five) are taken from the sample solution with a defined volume V (about 1 mL). They are spiked with a fixed small volume v (50 or 100 μL) of a standard solution (mostly aqueous solutions) containing the single analyte element in different concentrations $c_i (i = 0, 1,..., 4)$. Usually, the first concentration c_0 is chosen to be zero (ultrapure water), the second c_1 should be on the order of the unknown concentration. This value can roughly be estimated according to the respective peak height in the spectrum of the original solution. A certain amount of experience is of course valuable for such an estimation. The further concentrations c_2, c_3, and c_4 should be multiples (respectively about twofold, threefold, and fourfold) of the estimated value c_1. The final solutions should be thoroughly mixed to ensure homogeneity.

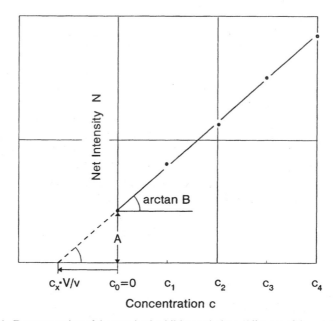

Figure 4-11. Demonstration of the standard-addition technique. Aliquots of the original solution (volume V) are spiked with a few standards (volume v) of increasing concentration $c_0,...., c_4$. The net intensities lead to a fitted straight line that can be extrapolated. Its intersection with the abscissa yields the unknown concentration c_x of the original sample multiplied by the factor V/v.

Finally, a certain volume w (about 10 μL) of the individual spiked solutions is used for recording the spectra. The net intensities N_i of the respective peaks of the analyte are plotted against the concentration c_i, as is demonstrated in Figure 4-11. Actually, the mass m_i of the analyte element taken by the aliquot w of the final solution should have been plotted on the abscissa. But both quantities c_i and m_i can be exchanged according to the definite relationship

$$c_i = m_i \frac{v + V}{vw} \tag{4-4}$$

While the applied mass m_i is on the order of ng, the concentration c_i of the standard solutions is on the order of μg/mL.

A linear relationship between net intensity N and mass m or concentration c can be expected because of the small amounts of the specimens. It is represented by a calibration straight line in Figure 4-11 and can be determined by a linear regression according to

$$N = A + Bc \tag{4-5}$$

where A is the ordinate offset, and B is the slope or absolute sensitivity. The unknown concentration c_x of the analyte can now be determined by an extrapolation of the straight line. It is calculated by the formula

$$c_x = \frac{A}{B} \frac{v}{V} \tag{4-6}$$

where the second fraction v/V is the dilution factor (1/10 to 1/20).

The error that is to be expected from this determination can be estimated from a confidence interval. If three or more additions are carried out $(n \geqslant 3)$ this interval will be on the order of

$$\Delta c = t(P, n - 2)\frac{s_R}{B}\frac{v}{V} \tag{4-7}$$

where t is the student factor for a chosen significance level P; n, the number of standard additions; s_R, the residual scatter of the measuring points around the straight line; and B, its slope.

This method of quantification is very easy, and relative sensitivity values are not required. Nevertheless, it is time consuming and therefore seldom used [38, 39]. It may be worthwhile when several more samples with the same matrix or composition have to be analyzed. In that case, the original straight line can be parallel shifted until it runs through the origin and can henceforth be used for external standardization. Furthermore, the standard addition technique is suitable for a later check of one result or the other of a multi-element determination.

4.3.2.2.   Multielement Determinations

The second method of internal standardization is based on the addition of an element initially not present in the sample. Generally, rare elements are chosen, e.g., Ga or Y in acidic solutions and Ge in basic solutions, possibly diluted with ultrapure water and ethanol. Medium-heavy elements with K-detection are favored over heavy elements with L-detection because of the minor number of peaks. Lighter elements with $Z \leqslant 21$ are not suited as standards because of particle-size effects that become troublesome in the low-energy range $E < 4$ keV.

As illustrated in Figure 4-12, a larger aliquot V of the sample (some mL) is taken and a smaller aliquot v of the standard (some μL) is added to it and thoroughly mixed. The added element henceforth serves as an internal standard for all other elements of the sample already detected and now to be quantitatively determined. For that purpose, a small volume w ($2-10\mu$L) of the final solution is pipetted on a clean carrier and dried by evaporation. The X-ray spectrum is recorded, providing the net intensities of detected elements within a preset live time. The concentration is then calculated by the simple formula

$$c_x = \frac{N_x/S_x}{N_{is}/S_{is}} c_{is} \tag{4-8}$$

where N is the net intensity; S, the relative sensitivity; and c, the concentration—of either analyte x or internal standard is, as indicated. This formula is independent of the volume w, but the concentration values are both related to the small volume v of the standard. However, since c_x is to be determined with respect to the original sample volume V, equation (4-8) must still be multiplied with the dilution factor v/V.

This method of internal standardization can be applied for solutions or suspensions and for samples that are prepared as solutions or suspensions. The standard is preferably added to the sample at an early stage of sample preparation and is homogeneously mixed. But it can even be added to a specimen already deposited on the sample carrier, although with the risk of an inhomogeneous distribution. For solid samples that are deposited as thin sections, the standard can only be added afterward [40]. The concentration

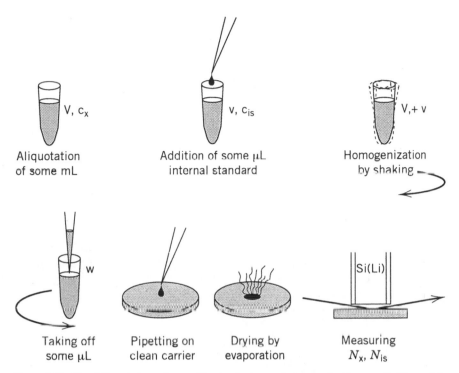

Figure 4-12. The different steps of quantification by internal standardization, applied for multi-element analyses by TXRF.

c_{is} of the internal standard is given by the quotient of the pipetted mass m_{is} and the total mass m_0 of the section. This total mass can be determined by a difference weighing with a microbalance. The concentration c_x of an analyte element again follows from equation (4-8).

Instead of the concentration c_x, the mass m_x has to be determined for airborne particulate matter deposited on a filter or on a carrier in a certain time t or from a certain volume V, respectively. The pipetted mass m_{is} of the internal standard must then be used instead of the concentration c_{is}, and equation (4-8) can be applied appropriately. For the analysis of contaminations, e.g., on a wafer surface, the mass m_x has to be related to the area limited by the detector's field of vision.

For microanalysis of powders, single grains, fibers, crumbs, or metallic smears deposited on the carriers, a weighing of the minute amount of micrograms is not possible. In that case, the concentration of elements with respect to this minute mass cannot be determined. But the *detected* elements can be quantified with respect to their sum c_0, which may be set to 100%.

A consequent modification of equation (4-8) results in the formula

$$c_x = \frac{N_x/S_x}{\sum\limits_{j} N_j/S_j} c_0 \qquad (4\text{-}9)$$

The summation must include all detected elements.

The reliability of multielement determinations can be estimated from the results obtained for multielement standards or from intercomparison tests. The precision of repeated determinations is about 1–5%, and the accuracy of the mean is only slightly poorer. Further details are given in Section 5.1.3.

4.3.3. Conditions and Limitations

Some conditions have to be fulfilled to enable the methods of quantification to be applied. The specimens must especially be limited in thickness and consequently in area-related mass or covering of the carrier. Both quantities should be restricted to a range with an upper and a lower limit each (d_{max} and d_{min} as well as m_{max} and m_{min}).

A first limitation is given by the geometry of excitation. The height of the primary beam is usually limited to about 20–30 μm, the height of the triangular region with a standing wave to about 10–15 μm. For that reason, the thickness of a specimen should be restricted to 15 μm.

The second reason for a restriction arises from the limited counting capability of the detector. A count rate of about 12,000 cps is a maximum value where the dead-time percentage D is 50%. This high count rate is achieved for a specimen amount mainly depending on the matrix. Upper values have been determined experimentally for three typical matrices characterized by their density ρ_{matr}: organic tissues (dried, ca. 0.2 g/cm^3), mineral powders (ca. 2 g/cm^3), metallic smears (ca. 8 g/cm^3). The upper limits m_{max} are 250 μg/cm^2 for organic tissues, 140 μg/cm^2 for mineral powders, and 8 μg/cm^2 for metallic smears. This is a rough estimate from unpublished measurements and can differ by a factor of 2 for the different excitation modes mentioned in Section 4.2.1.

The corresponding thickness d_{max} can be determined from m_{max} according to

$$d_{max} = m_{max}/\rho_{matr} \qquad (4\text{-}10)$$

provided that the carrier is uniformly covered. This quantity d_{max} is 12 μm for organic tissues, 0.7 μm for mineral powders, and 0.01 μm for metallic smears.

A further limitation is caused by X-ray absorption of the matrix for the analyte element. This absorption will remain below a certain permissible level, e.g., of 5%, if the thickness of the specimen is restricted to a maximum value d_{max}. This limit can be calculated by the formula

$$d_{max} = \frac{0.05/\rho_{matr}}{\overline{(\mu/\rho)}} \qquad (4\text{-}11)$$

$\overline{(\mu/\rho)}$ is a total mass-attenuation coefficient including the mass attenuation of the primary incoming beam and the secondary emerging beam, corrected for geometry:

$$\overline{(\mu/\rho)} = \sum c_i [(\mu/\rho)_{i,E_0}/\alpha + (\mu/\rho)_{i,E_x}] \qquad (4\text{-}12)$$

where values of c_i are the mass fractions of the constituents i of the matrix; E_0 is the energy of the primary photons; E_x, that of the fluorescence photons or peak of the analyte x, and α is the small glancing angle of incidence while the take off angle is assumed to be 90°. Respective values for d_{max} were calculated by Klockenkämper and von Bohlen [36]. Excitation mode b was provided, and K-$K\alpha$ or Cd-$L\alpha$ ($E \geqslant 3.1$ keV) was chosen as the fluorescence radiation. For the three aforementioned matrices, the upper limit was found to be 4, 0.05, and 0.002 μm, respectively.

This upper limit was deduced from the intensity of the analyte element alone, regardless of an internal standard. For that reason, the corresponding values are very restrictive. In practice, however, quantification is always performed by internal standardization. The net intensity of the analyte is related to that of an internal standard element in accord with equation (4-8). Variations of the two intensity values are largely compensated by the intensity ratio. The upper limit d_{max} can thus be deduced from the condition that the total matrix absorption differs by less than 5% for the analyte and internal standard. Condition (4-11) can then be replaced by the formula

$$d_{max} = \frac{0.05/\rho_{matr}}{|\overline{(\mu/\rho)}_x - \overline{(\mu/\rho)}_{is}|} \qquad (4\text{-}13)$$

It can be shown that the respective values are significantly above the limits determined by equation (4-11), generally by a factor of 10–400. The photon energy of the analyte and standard may even differ by threefold. But it has to be provided that the analyte and standard are homogeneously distributed in the specimen.

A lower limit for the area-related mass or covering is determined by the detection power, which sets a limit m_{min} of about 10 pg/cm^2. In addition, there is a lower limit for the thickness, determined by the modulated wave field above the carrier. This troublesome modulation can be leveled out if the specimen includes *several* nodes and antinodes. For that purpose, the thickness must exceed a certain minimum value d_{min}. It was determined by de Boer [41] for the three typical matrices to be 0.2, 0.05, and 0.015 μm, respectively.

Because of the internal standardization, the lower limit d_{min} can be reduced. Instead of several nodes and antinodes, only the first antinode above the carrier should contribute to excitation. The specimen should be as thick as the node distance given by equation (2-12). If the glancing angle α is chosen to be 70% of the critical angle α_{crit}, the lower limit is approximately

$$d_{min} \approx 22/\sqrt{\rho} \qquad (4\text{-}14)$$

The density ρ of the carrier is in g/cm^3; the thickness d_{min} is in nm. This quantity is independent of the excitation mode, the analyte, and the matrix. For quartz glass carriers, d_{min} is about 15 nm.

The decisive values for lower and upper limits of the covering and thickness of a specimen are summarized in Table 4-3 for the three typical matrices mentioned above. The lower limits are generally determined by the detection power and the height of the first antinode. The upper limits are set by the capability of the detector, i.e., by its maximum count rate. Matrix absorption is no longer a critical factor as internal standardization is carried out. The thickness of a specimen can even exceed the 10- or 100-fold of the tabulated value d_{max} if the covering m_{max} is kept constant. This can be achieved by an incomplete and nonuniform covering.

Even slighter restrictions may be necessary if the quantification is performed by standard additions. In this case, the analyte and internal standard are the same and the limitation by matrix absorption, expressed in equation

Table 4-3. Upper and Lower Limits of the Covering and Thickness of Three Typical Matrices

Matrices	Organic Tissues	Mineral Powders	Metallic Smears
Constituents	$_1$H...$_8$O	$_8$O...$_{20}$Ca	$_{24}$Cr...$_{30}$Zn
Covering:			
m_{max} (μg/cm^2)	250	140	8
m_{min} (μg/cm^2)	10^{-5}	10^{-5}	10^{-5}
Thickness:			
d_{max} (μm)	12	0.7	0.01
d_{min} (μm)	0.015	0.015	0.015

(4-13), becomes irrelevant. But also here, a homogeneous distribution of the analyte or standard is desirable in order to get accurate results.

4.4. QUANTITATIVE SURFACE AND THIN-LAYER ANALYSES

In addition to micro- and trace analyses, TXRF can be applied to surface and near-surface layer analyses. The samples to be examined have to be even and optically flat in order to ensure total reflection. This condition is largely fulfilled for wafers—flat and polished disks of, e.g., silicon or germanium. TXRF is highly suitable (i) for the detection of surface contaminations of such wafers, and (ii) for the characterization of near-surface layers, layer systems, or coatings of wafers. For that reason, TXRF plays an important role in the semiconductor industry.

The feasibility of thin-layer analysis is known from classical XRF and is based on continuous variation of the glancing angle of incidence while fluorescence intensity is recorded. The angle-dependent intensity profiles provide information on the elemental composition and thickness of surface layers. In contrast to the conventional method, however, TXRF is restricted to grazing incidence and occurs at total reflection. The glancing angle is only about 0.1° instead of some 10°, and the penetration depth is in the range of nanometers instead of micrometers. Furthermore, standing waves appear in front of a bulk sample or even within a layered structure. Their effect on the fluorescence intensity was dealt with in Section 2.3.

These specific features have to be considered for the recording and evaluation of angle-dependent intensity profiles of the samples. The shape and course of those curves allow a first qualitative evaluation of the samples which have to be examined. On the one hand, contaminations on or in a surface can be characterized and the covering or area-related mass of these contaminations can be determined. On the other hand, the elemental composition, the thickness, and even the density of near-surface layers can be evaluated. External standards of pure elements are used for calibration, and model calculations are carried out to fit the measured curves for quantification.

4.4.1. Recording Angle-Dependent Intensity Profiles

The basic design of an instrument suitable for surface and thin-layer analysis is shown in Figure 4-13. The primary beam is generated by a fine-focus X-ray tube, transmitted by a metal-foil filter, monochromatized by a multilayer or Bragg crystal, and directed onto the sample. Its fluorescence intensity is recorded by a Si(Li) detector, as was also shown in Figure 3-1. The specific components of the present arrangement are the monochromator, the wafer

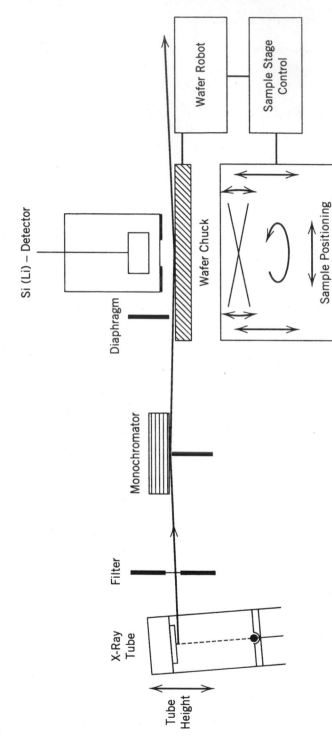

Figure 4-13. Schematic diagram of a TXRF instrument suitable for surface and thin-layer analyses of flat samples like wafers. All steps for positioning and measurement may be computer-controlled. To avoid contaminations, the equipment should be placed in a clean-bench.

robot, and the sample-positioning device. The entire instrument may be placed in a laminar flowbox in order to prevent contamination from ambient air.

A multilayer of Bragg crystal is needed to select a small energy band from the primary beam of a chosen X-ray tube, e.g., the $K\alpha$-peak of molybdenum or copper or the $L\beta$-peak of tungsten. The incident beam must be monochromatized in order to produce an unambiguous dependence of the intensity and the glancing angle. X-ray photons of *different* energies would blur this relationship. A quartz glass mirror acting as a low-pass filter can be employed for trace analyses with a fixed-angle adjustment but not for surface analyses with continuous angle variation. (Details on multilayers and Bragg crystals were given in Section 3.3.2).

The wafer robot may be used for loading and changing up to 50 wafers with a diameter of about 100–200 mm (4–8 in.). The wafers are placed automatically on a glass plate and cling to it by either an electrostatic or a vacuum chuck. Instruments are designed for 24 h unattended operation and measurement.

The sample-positioning device is the most important component (already described in Section 3.4.3). By its means, the sample or wafer is first adjusted in the observation plane. Mechanical sensors should not be employed in order to avoid any physical contact with the sample. The adjustment can simply be controlled by the fluorescence signal generated by the sample itself. Thereafter, the sample or wafer is tilted stepwise in order to run an angle scan. Starting from a setting close to zero, the incident angle is increased stepwise and fluorescence spectra are recorded during intervals between steps. The step width may be chosen between 0.005° and 0.05°, the counting time, between 50 and 500 s. About 10–50 steps are taken to reach an angle of about twice the critical angle of the sample material, which is generally < 1°. The total measuring time may be between 15 min and 4 h. After that, the peak intensities of those elements that have been detected are plotted against the incident angle. These angle-dependent intensity profiles are used for a qualitative and quantitative characterization of the samples under investigation.

The angle scale *can* be calibrated by taking photographs of the primary beam and the reflected beam. But it is much easier to determine the critical angles for some reference samples. For this purpose, polished wafers, disks of metals, or correspondingly plated wafers are suitable [42, 43]. The intensity profiles of these elements are recorded, and the inflection points of these curves (similar to that of Figure 2-14) are determined. They are set equal to the values calculated by equation (1-34). A correction of an additive constant may still be necessary. It can be determined by a second measurement after a horizontal rotation of the sample by 180°. This calibration technique [42] allows an angle setting with an absolute accuracy of about 0.005°. A divergence of the primary beam of about 0.01° is twice this value.

4.4.2. Distinguishing Between Types of Contamination

In the region of total reflection, the fluorescence intensity is strongly dependent on the glancing angle of incidence. From Section 2.3 we know that some
special kinds of samples show a certain typical dependence which may be
characteristic. Three of them will now be examined in detail: infinitely thick
and flat substrates, granular residues *on* a subrate, and buried layers *in*
a substrate. This inspection aims at the determination of trace impurities either
homogeneously distributed in thick samples, located in granular particles, or
evenly deposited in thin near-surface layers. TXRF is capable of such a determination and especially of distinguishing among these three contaminations,
which may be called *bulk type*, *particulate type*, and *thin-layer type* [44, 45].

To ensure total reflection, the sample or substrate must be optically flat and
even. Polished wafers designated for integrated circuits are ideally suited as
samples, as long as they are not patterned. Moreover, the control of their
contamination is highly important and has made TXRF an indispensable
analytical tool for the semiconductor industry.

The three different angle-dependent intensity profiles are demonstrated in

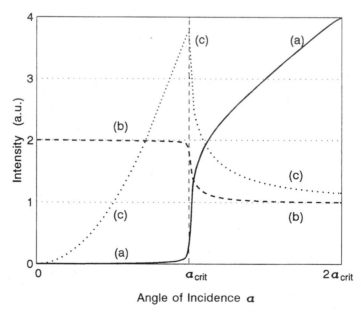

Figure 4-14. Characteristic intensity profiles for three different kinds of contaminations: (a) bulk
type; (b) particulate type; and (c) thin-layer type. The critical angle α_{crit} is determined by total
reflection at the flat substrate. (After Weisbrod et al. [45].)

Figure 4-14. The fluorescence intensity of some impurity element is plotted against the glancing angle α. Curve (a) is valid for a bulk-type contamination. The intensity is very small *below* the critical angle of the substrate but increases rapidly *above* this angle in accord with equation (2-27). It can be written in a modified notation as

$$N_B(\alpha) = c_{VB}[1 - R(\alpha)]\alpha N_{OB} \qquad (4\text{-}15)$$

where N_B is the net intensity of the analyte; c_{VB}, the volume-related concentration in atoms/cm^3; R, the reflectivity of the substrate given by equation (1-41); α, the glancing angle of incidence; and N_{OB}, a reference value used for normalization at a larger angle. Such a profile will be recorded if impurities are homogeneously distributed within the substrate. With increasing angle α, the penetration and information depth increases from several nanometers up to several micrometers.

For a particulate-type contamination, i.e., for curve (b), the intensity *above* the critical angle is nearly constant, since the particles are completely excited independent of the angle of incidence. Because of the total reflection, the intensity doubles at the critical angle in nearly steplike fashion and remains at the twofold value down to very small angles. This dependence already expressed by equation (2-21) can be written as

$$N_P(\alpha) = c_{AP}[1 + R(\alpha)]N_{OP} \qquad (4\text{-}16)$$

where c_{AP} is the area-related concentration in atoms/cm^2, and N_{OP} is a second reference value. This profile can be recorded if one or more grains of a certain size are located on a substrate. Their diameter has to be some 100 nm at a minimum and several micrometers at a maximum. But also a total set of grains is permissible that may have a Gaussian or a logarithmic Gaussian distribution of the grain size with an appropriate width [41].

For a thin-layer-type contamination, curve (c) is valid. The intensity far *above* the critical angle is constant, just as for curve (b). The asymptotic intensities of both curves will even be equal if both concentration values c_{AL} and c_{AP} are the same. Obviously, the particulate and thin-layer types do not differ in fluorescence at larger angles. But the intensity of curve (c) steadily increases to a three- to fourfold value just *above* the critical angle and thereafter decreases to zero in a parabolic curve. This profile can be described by a modified version of equation (2-23):

$$N_L(\alpha) = c_{AL}[1 - R(\alpha)]\frac{\alpha}{z_n(\alpha)}N_{OL} \qquad (4\text{-}17)$$

where c_{AL} is the area-related concentration in atoms/cm^2; z_n, the penetration depth of the primary beam in the substrate, normal to its surface, which is given by equation (1-43); and N_{OL}, a reference value that must be equal to N_{OP} of equation (4-16). Such a profile will only be recorded if an ultrathin layer of ca. 1 nm is deposited *above* or embedded *below* the surface of a thick substrate. No further differentiation of these two layers is possible [46].

It can be concluded from the preceding discussion that each of the three contaminations can be identified by its characteristic profile. Even a mixture of two or three types of contamination can be analyzed and the different fractions be determined [44]. For that purpose, an angle scan has to be performed and intensity profiles should be plotted for all elements under consideration. Such a profile is shown in Figure 4-15 as an example. It can directly be concluded from this profile that all three types of contamination are present.

In order to obtain quantitative results, the intensity sum is derived from equation (4-15), (4-16), and (4-17) and fitted to the measured curve. The concentration values c_{VB} for the bulk-type, c_{AP} for the particulate-type, and c_{AL} for the thin-layer-type contamination are varied until a best fit is reached. The ultimate values are treated as the results.

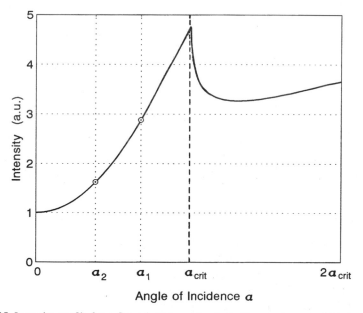

Figure 4-15. Intensity profile for a flat substrate contaminated by an element which is as well embedded throughout the substrate, as located in small grains above the substrate, as deposited as an ultrathin layer on top of or just below the subtrate. The critical angle α_{crit} is determined only by the substrate itself; α_1 and α_2 are significant angles of operation.

The method, however, must first be calibrated. This task consists in determining the calibration factors N_{OB} and N_{OP} or N_{OL}. These quantities essentially are a product of the relative sensitivity of the analyte and the intensity of the primary beam. They are further determined by the total mass-attenuation coefficient of the matrices in accord with equation (4-12). Calibration by internal standardization is not permissible because any additional contamination should be avoided here. Consequently, an external calibration becomes necessary, to be carried out by means of an external standard. An overview of different methods is given by Hockett [47]. Some users have applied a particulate-type standard presented by a microdroplet of a standard solution after drying [48]. Other users have recommended a layer-type standard produced by immersion and/or spin-drying of the wafer with a spiked solution [49, 50]. Others have used bulk-type standards in the form of polished pure metals or plated wafers [42,51].

All types of standards are commercially available, but it is strongly recommended that their intensity profiles be inspected prior to use [51]. Standards that do not show the appropriate profile should be rejected. For practical work, only one external-standard element is necessary. Measurement at one appropriate angular position gives the calibration factors N_{OB}, N_{OP}, or N_{OL} for the given element after application of equations (4-15), (4-16), or (4-17), respectively. Calibration factors for other elements can be deduced from this value by relative sensitivities.

The determination of contaminants and their differentiation by recording a total intensity profile is rather time consuming and not suitable for mapping of an entire wafer. There is, however, a fast method that limits the measurements to one or two angular positions [43,44]. If only the sum of a particulate- and layer-type contamination has to be determined, a single measurement will suffice. The appropriate angle of operation can be found at the intersection of the two curves (b) and (c) in Figure 4-14. It can be derived by comparing equations (2-24) and (2-21). Assuming a reflectivity R of nearly 100%, we find that the angle of operation is approximately

$$\alpha_1 = \alpha_{crit}/\sqrt{2} \qquad (4\text{-}18)$$

This is the only angle below α_{crit} where the intensity sum is proportional to the concentration sum of both contaminations. This sum $c_{AL} + c_{AP}$ can be directly determined from the intensity reading N_1 at the angle α_1:

$$c_{AL} + c_{AP} = 0.5\, N_1/N_{OL} \qquad (4\text{-}19)$$

If the percentage of the particulate or the layer type has to be determined as well, a second measurement will be necessary. It may be carried out at a smaller angle of operation, e.g.,

$$\alpha_2 = \alpha_{crit}/\sqrt{6} \tag{4-20}$$

Both angles of operation are indicated in Figure 4-15. The respective intensity readings N_1 and N_2 lead to the percentage c_L of the layer type defined by the ratio $c_{AL}/(c_{AL} + c_{AP})$. From equation (2-24) it can be shown that this quantity is given by

$$c_L = \frac{3}{2}\left(1 - \frac{N_2}{N_1}\right) \cdot 100\% \tag{4-21}$$

The corresponding percentage of the particulate type c_P defined by the ratio $c_{AP}/(c_{AL} + c_{AP})$ is given by

$$c_P = \frac{3}{2}\left(\frac{N_2}{N_1} - \frac{1}{3}\right) \cdot 100\% \tag{4-22}$$

Both values sum up to 100%. The calculations presuppose that samples or wafers are ideally flat and even. Otherwise, the results only represent approximate values.

4.4.3. Characterization of Thin Layers[1]

Thin-layered materials with a layer thickness in the range of nanometers are used as high-tech materials especially in integrated circuit (IC) technology and the glass industry. Several X-ray techniques are suitable for their characterization provided that grazing incidence is achieved. Glancing angles in the range of total reflection are necessary so as to be surface sensitive. TXRF is the most comprehensive method of these techniques, which makes compositional depth profiling possible, even nondestructively. To do this, the layers and the substrate certainly must be homogeneous, flat, and even, one by one. In other words, the compositional profile along the depth should be steplike, with perfectly sharp interfaces. Layers with a roughness of only about 1 nm and a curvature of some 100 m diameter are ideally capable of total reflection. For these stratified structures, TXRF can yield the element composition, the thickness, and the density of the individual layers with an accuracy of a few percent. With some mathematical effort, even a certain roughness of the interfaces can be taken into account.

---

[1] The term *thin film* is avoided here since it should be restricted to liquid layers produced from the liquid phase.

The record of angle-dependent intensity profiles is the basis for the evaluation. These profiles represent values of fluorescence intensities at different angle positions. The intensities are measured simultaneously for all elements of the layered sample at each angle position. The glancing angle is subsequently increased between about 0° and 0.5° or even 1° in steps of about 0.01°. From these profiles, the respective layer parameters are evaluated by an iterative fitting procedure on the basis of modeling calculations. It was developed and described independently by Weisbrod et al. [52] (see also Schwenke et al. [53]) and by de Boer [54], and contains five steps.

i. First of all, a starting model should be drawn up. By a qualitative interpretation of the element profiles, the number and sequence of the individual layers, their element composition, thickness, and density can roughly be guessed. For that purpose, a qualitative understanding of the course and the oscillation structure of the profiles is needed. Some hints can be found in Figure 4-16 a–f, mostly based on the findings presented in Chapter 2.

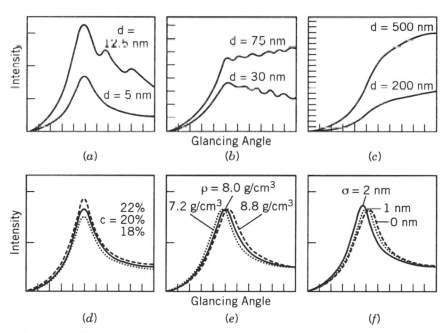

Figure 4-16. Different types of intensity profiles distinguished by four parameters of a model layer with a weight fraction $c = 20\%$ for a given element, a thickness $d = 5$ nm, a density $\rho = 8$ g/cm^3, and a roughness $\sigma = 0$ nm. In parts (a), (b), and (c), the thickness d is increased by about the 2.5-fold of the previous value; in (d) only c is varied by 10%; in (e) ρ is varied while ρd is kept constant; in (f) a roughness of 1 and 2 nm is implemented (after Hüppauf [57]; see also last paragraph)

ii. The next step involves calculation of the intensity of the primary beam within the layer system proposed by the starting model. This primary intensity is calculated either by a matrix formalism (as described in Section 2.4) or by a recursion formalism. The intensity I_0 is determined for a given photon energy E_0 and for different glancing angles α_i dependent on the depth z normal to the surface. It reflects the inhomogeneous wave penetrating the layer system as an evanescent or standing wave with nodes and antinodes parallel to the surface.

iii. The fluorescence intensity of elements excited by the primary beam is calculated next. For this purpose, a fundamental parameter approach is employed. The absorption of the fluorescence radiation by the layer system and integration over the depth z is a must and can be performed by means of the closed expression of equation (2-51). Secondary excitation or enhancement can also be taken into account [52,54], but as a complicated second-order effect it may be neglected for layers up to 100 nm thickness. The efficiency ε of the detector (see Section 3.5.3.1) and the transmission T of ambient air between the sample and the detector, however, can be included quite easily. To get intensity values really indicated by the detector, the angular divergence of the instrumentation may finally be taken into consideration. The calculated intensity I of the analyte element x may be written [53] as

$$I_x = I_x[(c_{x1}, d_1, \rho_1), \ldots, (c_{x, N+1}, d_{N+1}, \rho_{N+1}); \alpha, \varepsilon_x, T_x, K] \qquad (4\text{-}23)$$

where $c_{x1}, \ldots, c_{x, N+1}$ is the concentration of element x in the successive layers $1, \ldots, N+1$; $d_1, \ldots, d_{N+1}$ is the thickness and $\rho_1, \ldots, \rho_{N+1}$ is the density of these layers; α is the glancing angle of the primary beam; and K is a calibration constant that may be determined from a bulk sample [42]. The $(N+1)$th layer usually is the substrate.

iv. The calculated curve $I_x(\alpha)$ is compared with the intensity profile experimentally recorded. The residual scatter, i.e., the relative standard deviation between measured and calculated intensities, is actually determined. If the deviation is larger than a given value of about 5%, the starting model will be changed. Figure 4-16 may again be used as a guide for modifications. Steps **ii** to **iv** are then repeated in an iterative process until a best fit with a sufficiently small residual scatter is achieved.

v. A best fit is usually reached after about four to eight iterations. The last model with parameters $c_{x1}, \ldots, c_{x, N+1}$; $d_1, \ldots, d_{N+1}$; and $\rho_1, \ldots, \rho_{N+1}$ is declared to be the final result.

A few software programs have been developed on this subject. They require some sets of fundamental parameters such as photoelectric mass-absorption coefficients in an energy range from 1 to 60 keV, absorption-edge jump factors, fluorescence yields, relative emission rates for the individual peaks in their series, absorption edges, peak energies, as well as densities, atomic number, and atomic mass of pure elements. Respective sets of data are available. Nevertheless, the problem is rather complex and requires a skilled operator with considerable experience in interpretation of the profiles and modification of the models. Only thus can the iterative process converge rapidly and yield an unambigous result.

In practice, perfectly flat surfaces and interfaces are non-existent. A certain nanometric roughness will be left on bulk samples even after careful polishing. Indeed, an average roughness of some 10 to a few 100 nm often occurs in practice. Various approaches have sought to cope with such rough surfaces [55–58]. The implementation of a multilayer model first used by Nevot et al. [59,60] seems to be rather successful. It describes the zone of roughness as a stack of virtual planeparallel layers with a stepwise increasing density [55,57]. The uppermost layer may show zero density, while the lowest layer should have the bulk density. This subject, however, is still in the developmental stage and goes beyond the scope of this presentation.

REFERENCES

1. IUPAC Nomenclature (Revision by Sandell, E.B., West, T.S., Flaschka, H., and Menis, O.) (1979). *Pure Appl. Chem.* **51**, 43.
2. Jenkins, R. (1988). *X-Ray Fluorescence Spectrometry*, Chemical Analysis Series. Wiley (Interscience), New York.
3. Tschöpel, P., Kotz, L., Schulz, W., Veber, M., and Tölg, G. (1980). *Fresenius' Z. Anal. Chem.* **302**, 1.
4. Prange, A. (1994). *Labor Praxis* **18/2**, 30.
5. Knoth, J., and Schwenke, H. (1977). German patent pending No. P 2717925.5.
6. Michaelis, W., Knoth, J., Prange, A., and Schwenke, H. (1985). *Adv. X-Ray Anal.* **28**, 75.
7. von Bohlen, A., Eller, R., Klockenkämper, R., and Tölg, G. (1987). *Anal. Chem.* **59**, 2551.
8. Bilbrey, D.B., Leland, D.J., Leyden, D.E., Wobrauschek, P., and Aiginger, H. (1987). *X-Ray Spectrom.* **16**, 161.
9. Leland, D.J., Bilbrey, D.B., Leyden, D.E., Wobrauschek, P., Aiginger, H., and Puxbaum, H. (1987). *Anal. Chem.* **59**, 1911.
10. Salvà, A., von Bohlen, A., Klockenkämper, R., and Klockow, D. (1993). *Quimi. Anal.* **12**, 57.
11. Ketelsen, P., and Knöchel, A. (1984). *Fresenius' Z. Anal. Chem.* **317**, 333.

12. Knöchel, A., Bethel, U., and Hamm, V. (1989). *Fresenius' Z. Anal. Chem.* **334**, 673.

13. Reus, U. (1991). *Spectrochim. Acta* **46B**, 1403.

14. Prange, A. (1987). *GIT Fachz. Lab.* **6**, 513.

15. Prange, A., Knoth, J., Stössel, R.P., Böddeker, H., and Kramer, K. (1987). *Anal. Chim. Acta* **195**, 275.

16. Prange, A. (1989). *Spectrochim. Acta* **44B**, 437.

17. Prange, A., Böddeker, H., and Michaelis, W. (1990). *Fresenius' Z. Anal. Chem.* **335**, 914.

18. Gerwinski, W., and Goetz, D. (1987). *Fresenius' Z. Anal. Chem.* **327**, 690.

19. Knöchel, A., Dierks, H., Hastenteufel, S., and Haurand, M. (1989). *Fresenius' Z. Anal. Chem.* **334**, 673.

20. Stössel, R.P., and Prange, A. (1985). *Anal. Chem.* **57**, 2880.

21. Reus, U., Freitag, K., and Fleischhauer J. (1989). *Fresenius' Z. Anal. Chem.* **334**, 674.

22. Reus, U. (1989). *Spectrochim. Acta* **44B**, 533.

23. Chen, J.S., Berndt, H., Klockenkämper, R., and Tölg, G. (1990) *Fresenius' Z. Anal. Chem.* **338**, 891.

24. Prange, A., Knöchel, A., and Michaelis, W. (1985). *Anal. Chim. Acta* **172**, 79.

25. Burba, P., Willmer, P.G., Becker, M., and Klockenkämper, R. (1989). *Spectrochim. Acta* **44B**, 525.

26. Knoth, J., and Schwenke, H. (1978). *Fresenius' Z. Anal. Chem.* **291**, 200.

27. Knoth, J., and Schwenke, H. (1979). *Fresenius' Z. Anal. Chem.* **294**, 273.

28. Klockenkämper, R., von Bohlen, A., Moens, L., and Devos, W. (1993). *Spectrochim. Acta* **48B**, 239.

29. Schneider, B. (1989). *Spectrochim. Acta* **44B**, 519.

30. Bredendiek-Kämper, S., von Bohlen, A., Klockenkämper, R., Quentmeier, A., and Klockow, D. (1996). To be published in *J. Anal. At. Spectrom.*

31. von Bohlen, A., Klockenkämper, R., Tölg, G., and Wiecken, B. (1988). *Fresenius' Z. Anal. Chem.* **331**, 454.

32. Knoth, J., Schneider, H., and Schwenke, H. (1994). *X-ray Spectrom.* **23**, 261.

33. Rigaku Industrial Corporation (1992). *Rigaku J.* **9**, 29 (tech. note).

34. Bearden, J.A. (1964). "X-ray Wavelengths," Rep. NYO-10586. U.S. Atomic Energy Commission, Washington, D.C.

35. Johnson, G.G., Jr., and White, E.W. (1970). *X-ray Emission Wavelengths and keV Tables for Nondiffractive Analysis*, ASTM Data Series DS 46. American Society for Testing Materials, Philadelphia.

36. Klockenkämper, R., and von Bohlen, A. (1989). *Spectrochim. Acta* **44B**, 461.

37. Welz, B. (1983). *Atomabsorptionsspektrometrie*, 3rd ed., pp. 121–123. Verlag Chemie, Weinheim.

38. Ninomiya, T., Nomura, S., Taniguchi, K., and Ikeda, S. (1989). *Adv. X-ray Anal.* **32**, 197.

39. Yap, C.T. (1988). *Appl. Spectrosc.* **42**, 1250.

40. Klockenkämper, R., von Bohlen, A., and Wiecken, B. (1989). *Spectrochim. Acta* **44B**, 511.

41. de Boer, D.K.G. (1991). *Spectrochim. Acta* **46B**, 1433.

42. Gutschke, R. (1991). Diploma thesis, University of Hamburg.

43. Berneike, W. (1993). *Spectrochim. Acta* **48B**, 269.

44. Schwenke, H., and Knoth, J. (1995). *Part. Surf.* [*Proc. Symp.*], *Meet., 1992*, pp. 311–323.

45. Weisbrod, U., Gutschke, R., Knoth, J., and Schwenke, H. (1991). *Fresenius' J. Anal. Chem.* **341**, 83.

46. de Boer, D.K.G., and van den Hoogenhof, W.W. (1991). *Spectrochim. Acta* **46B**, 1323.

47. Hockett, R.S. (1995). *Adv. X-ray Chem. Anal. Jpn.* **26s**, 79.

48. Fabry, L., Pahlke, S., Kotz, L., Adachi, Y., and Furukawa, S. (1995). *Adv. X-ray Chem. Anal. Jpn.* **26s**, 19.

49. Torcheux, L., Degraeve, B., Mayeux, A., and Delamar, M. (1994). *SIAJ.* **21**, 192.

50. Mori, Y., Shimanoe, K., and Sakon, T. (1995). *Adv. X-ray Chem. Anal. Jpn.* **26s**, 69.

51. Schwenke, H., and Knoth, J. (1995). *Adv. X-ray Chem. Anal. Jpn.* **26s**, 137.

52. Weisbrod, U., Gutschke, R., Knoth, J., and Schwenke, H. (1991). *Appl. Phys. A* **53**, 449.

53. Schwenke, H., Gutschke, R., and Knoth, J. (1992). *Adv. X-ray Anal.* **35B**, 941.

54. de Boer, D.K.G. (1991). *Phys. Rev. B* **44**, 498.

55. Schwenke, H., Gutschke, R., Knoth, J., and Kock, M. (1992). *Appl. Phys. A* **54**, 460.

56. van den Hoogenhof, W.W., and de Boer, D.K.G. (1993). *Spectrochim. Acta* **48B**, 277.

57. Hüppauf, M. (1993). Doctoral thesis, RWTH Aachen, and JÜL-report JÜL-2730, ISSN 0366-0885.

58. Kawamura, T., and Takenake, H. (1994). *J. Appl. Phys.* **75**, 3806.

59. Névot, L., and Croce, P. (1980). *Rev. Phys. Appl.* **15**, 761.

60. Névot, L., Pardo, B., and Corno, J. (1988). *Rev. Phys. Appl.* **23**, 1675.

CHAPTER

5

EFFICIENCY AND APPLICABILITY OF TXRF

As regards its efficiency, the special analytical method of TXRF may be characterized as being relatively economical with respect to costs of instrumentation and maintenance, as well as low in time consumption. Simultaneous multielement detection is possible, with detection limits in the low-pg range. Quantitative determinations can be carried out by the simple and reliable method of internal standardization. Moreover, TXRF is applicable to a great variety of sample materials. Samples of environmental or biological origin are often analyzed for monitoring purposes when large numbers of items must be dealt with. Medical or clinical material is frequently investigated by TXRF, as quite often only small sample amounts are available. The microcapability of TXRF is also useful to help resolve art-historical or forensic questions when precious works of art or unique pieces of evidence have to be analyzed. A new field of applications has opened up in the semiconductor industry, where TXRF is used for surface and thin-layer analyses of wafers. Because of its efficiency and competitiveness, TXRF will no doubt play an increasingly important role within the family of atomic spectrometric methods.

5.1. ANALYTICAL CONSIDERATIONS

The efficiency of TXRF is first of all a result of the fact that it is an energy-dispersive method of X-ray fluorescence analysis. Specimens need be deposited on totally reflecting carriers in only small amounts. This capability for microanalyses is the second feature that determines the efficiency of TXRF. The benefits of this efficiency affect costs, detection power, reliability, and applicability.

5.1.1. General Costs of Installation and Upkeep

The simplicity of the TXRF instrumentation described in Chapter 3 ought to be reemphasized. Complete instruments as well as individual supplements or components are commercially available. A number of manufacturers offer a certain selection of devices. The basic purchase costs amount to about

173

$200,000[1] for a complete instrument including a high-voltage (HV) generator, X-ray tubes, reflectors, sample changer, detector, multichannel analyzer, computer, and software. An instrument additionally offering an angle scan of the sample and therefore suited for surface and thin-layer analyses may cost some $50,000 more. The complete instruments are already adjusted, mechanically stable, and nearly maintenance free. Operational problems are very infrequent, but sometimes a leakage of the thin detector window does occur, which then has to be replaced by the manufacturer.

For installation, a space of only about $4 \, m^2$ is required, and the room should be air-conditioned. A power supply of 3 kW is needed, and a coolant system with a flow rate of 5 L/min must be connected. The detector needs to be cooled with liquid nitrogen, and an accessory Dewar of 10 L volume has to be filled at least once but not more than twice a week. A vacuum will not be needed if one does not intend to undertake the detection of light elements with low atomic numbers ($Z < 14$). But for light-element detection a standard vacuum (1 hPa) is necessary; furthermore, a detector with a special diamond-like window or even a windowless detector will have to be utilized, the later of which may be more susceptible to trouble.

Sample preparation as a first step before analysis takes the bulk of the time and effort, especially when samples cannot be analyzed directly but have to be digested prior to analysis. However, a lot of preparation techniques are well tried and tested for approved methods like ET-AAS or ICP-OES and can also be applied for TXRF. For analysis, a small specimen of about $10 \, \mu L$ or $10 \, \mu g$ of a sample has to be deposited on a flat glass carrier and dried by evaporation. Sample preparation and presentation necessitate an extremely clean working space, preferably on a clean bench of class 100[2]. The glass carriers chosen for deposition should first be checked for cleanliness. The expensive quartz glass carriers ($28 each) can be cleaned and reused; the cheap Plexiglas carriers (10 ¢ each) are used only once.

The next step for TXRF analysis is the recording of an energy-dispersive spectrum, which is rather straightforward. A total spectrum can be recorded in some seconds, but usually a counting time of 1 or 2 min is chosen and for extreme traces 20 min are preset. The total spectrum is recorded simultaneously, so that even an element that may be in the sample or specimen unexpectedly will be detected and no element will be overlooked. The processing of the spectra is likewise simple and rapid, and is usually done via a software program. Quantification is performed by internal standardization, either by addition of the analyte element in a few concentration steps or by addition of

---

[1] All costs are given in U.S. currency here.

[2] Class 100 for a clean room means: < 100 particles/cft or $< 3500/m^3$ with diameters $\geqslant 0.5 \, \mu m$; no particles $\geqslant 5 \, \mu m$ (after U.S. Fed. Stand. 209 D).

one other element (i.e., previously not present in the sample) in a single concentration level. The latter method, which is normally applied for multiele-ment determinations, saves a lot of time and effort (see Section 4.3.2).

For micro- and trace analyses, only one angle position is used. For surface and (even more) for thin-layer analyses, the sample must be tilted in several steps for an angle scan. The time needed for angle-dependent intensity profiles can extend to about 1 h or more. The evaluation requires a correction of absorption-enhancement effects by a fundamental parameter method. An appropriate software program is needed for the complex calculations.

5.1.2. Detection Power for Elements

TXRF is especially valuable because of its high detection power. All elements with atomic numbers $Z \geqslant 11$ (sodium) can be detected without the need for a vacuum. If windowless detectors are used and a vacuum is applied, even the lighter elements down to $Z = 6$ (carbon) may be detectable and the other elements with atomic numbers $Z \leqslant 20$ (calcium) can be determined at a higher sensitivity. Elements can be detected simultaneously if their spectral peaks do not strongly overlap, but strong interference is rather an exception. Generally, some 10–15 (but anyhow less than 20) different elements can be determined in a single run. In order to detect all potential elements of a sample, two or even three runs are needed at different excitation modes, as mentioned in Sec-tion 4.2. Consequently, two or three spectra have to be recorded and evaluated for the same specimen.

Detection limits can be determined according to IUPAC rules [1]. The minimum detectable amount or mass m_{min} of an element can be calculated from the formula

$$m_{min} = k \frac{s_{blank}}{B} \qquad (5\text{-}1)$$

where k is a factor for which a value of 3 is strongly recommended; s_{blank}, the standard deviation of the blank measurement; and B, the absolute sensitivity for the element of interest. For X-ray spectroscopical methods, s_{blank} can widely be influenced by photon counting of the spectral background. It is limited by Poisson statistics:

$$s_{blank} \geqslant \sqrt{\bar{N}_{back}} \qquad (5\text{-}2)$$

where $\bar{N}_{back}$ is the averaged background. If the blank or background correc-tion is carried out separately for each single spectrum and not for the averaged blank spectrum, $\sqrt{(2N_{back})}$ is the decisive quantity. The sensitivity B can be measured by the net counts N_{net} of the analyte element related to the quantity m of this element coming from the specimen. Consequently, an ideal value of the detection limit can be determined from

$$m_{min} = 3 \frac{m}{N_{net}} \sqrt{(2 N_{back})} \qquad (5\text{-}3)$$

Detection limits were measured with the residues of aqueous standard solutions [2, 3] and are presented in Figure 5-1. The three excitation modes already described in Section 4.2 were applied. The live time was chosen to be 1000 s, so that the actual elapsed time was about 1200 s (20 min) due to a dead-time portion D of 20%. The Si(Li) detector used for these measurements had a frontal area of 80 mm^2 and a spectral resolution of 150 eV for the Mn-$K\alpha$ peak.

Figure 5-1 shows the familiar relationship between the detection limit and the atomic number with a minimum for a particular analyte. All elements with atomic numbers $Z \geq 11$ can be detected either by their K-peaks or L-peaks.

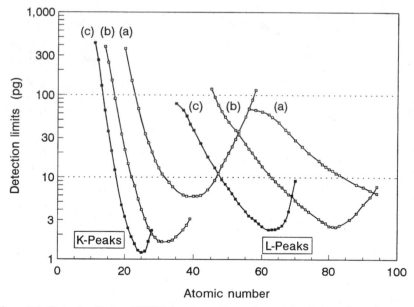

Figure 5-1. Detection limits of TXRF for the residues of aqueous solutions, dependent on the atomic number of the analyte element. Three excitation modes were used: (a) W-tube, 50 kV; Ni-filter, cutoff 35 keV; (b) Mo-tube, 50 kV; Mo-filter, cutoff 20 keV; (c) W-tube, 25 kV; Cu-filter. The three curves to the left were determined by the detection of K-peaks; the three curves to the right, by that of L-peaks. (After Reus et al. [2] and Prange and Schwenke [3].)

The detection limits go down to the low-picogram level. By means of an appropriate excitation, *nearly all* elements are detectable at a level of 1–10 pg. Consequently, the ng/L region would be reached if 100 μL of a high-purity water or acid were applied, and the ng/g range would be reached if 100 μg of an organic matrix were used. Metal contaminants on wafer surfaces could thus be determined down to or even below 10^{10} atoms/cm^2.

Such ideal values, however, can only be achieved for minute residues of aqueous solutions. It is a further prerequisite that the main spectral peaks of the analytes be free from interferences of a further constituent. Finally, special blanks must not disturb analysis and only photon fluctuations should be significant for the background noise. In practical examples, these conditions are often only partly fulfilled.

Relative detection limits have actually been determined in many practical applications. These real values collected from the literature are represented in Figure 5-2. The actual detection limits are dependent on the original sample matrix and the kind of sample preparation or matrix separation. For high-purity waters or acids, detection limits go down to the low-μg/L level after simple drying of μL specimens. After freeze-drying of mL volumes, detection limits are lowered to the ng/L level. For natural waters burdened with dissolved or suspended matter, digestion and matrix separation are necessary to achieve detection limits below μg/L. For direct analysis of light samples with a biomedical or environmental origin, detection limits are at about 0.1 μg/g but can be improved to 10 ng/g if the matrix is first digested. Inorganic solid samples can be analyzed down to about 0.1 μg/g after digestion and matrix separation, e.g., high-purity metals.

5.1.3. Reliability of Determinations

Because the spectrum is recorded as a whole, no detectable element may be overlooked, but because of a low spectral resolution, some peak interferences can occur. For example, the $K\alpha$-peaks of the transition metals $_{21}$Sc to $_{30}$Zn are partly overlapped by the $K\beta$-peaks of the neighboring elements with lower atomic number. The $L\alpha$ and $L\beta$-peaks of the lanthanoids overlap in a similar way. Furthermore, additional peaks like escape or sum peaks can occur that have to be noticed carefully (Sections 3.5.3.4 and 4.2.3.1). In general, however, the identification of the low-numbered peaks in an X-ray spectrum is easy, the detection of elements is unambiguous, and a qualitative analysis can be carried out rapidly and reliably.

For quantitative analysis, the determination of the net intensity of peaks is relevant, ranging from about 10^2 to some 10^6 counts. For that reason, the determination of the content of *one* element is restricted to 4 orders of magnitude. This dynamic range can be spanned for one definite instrumental

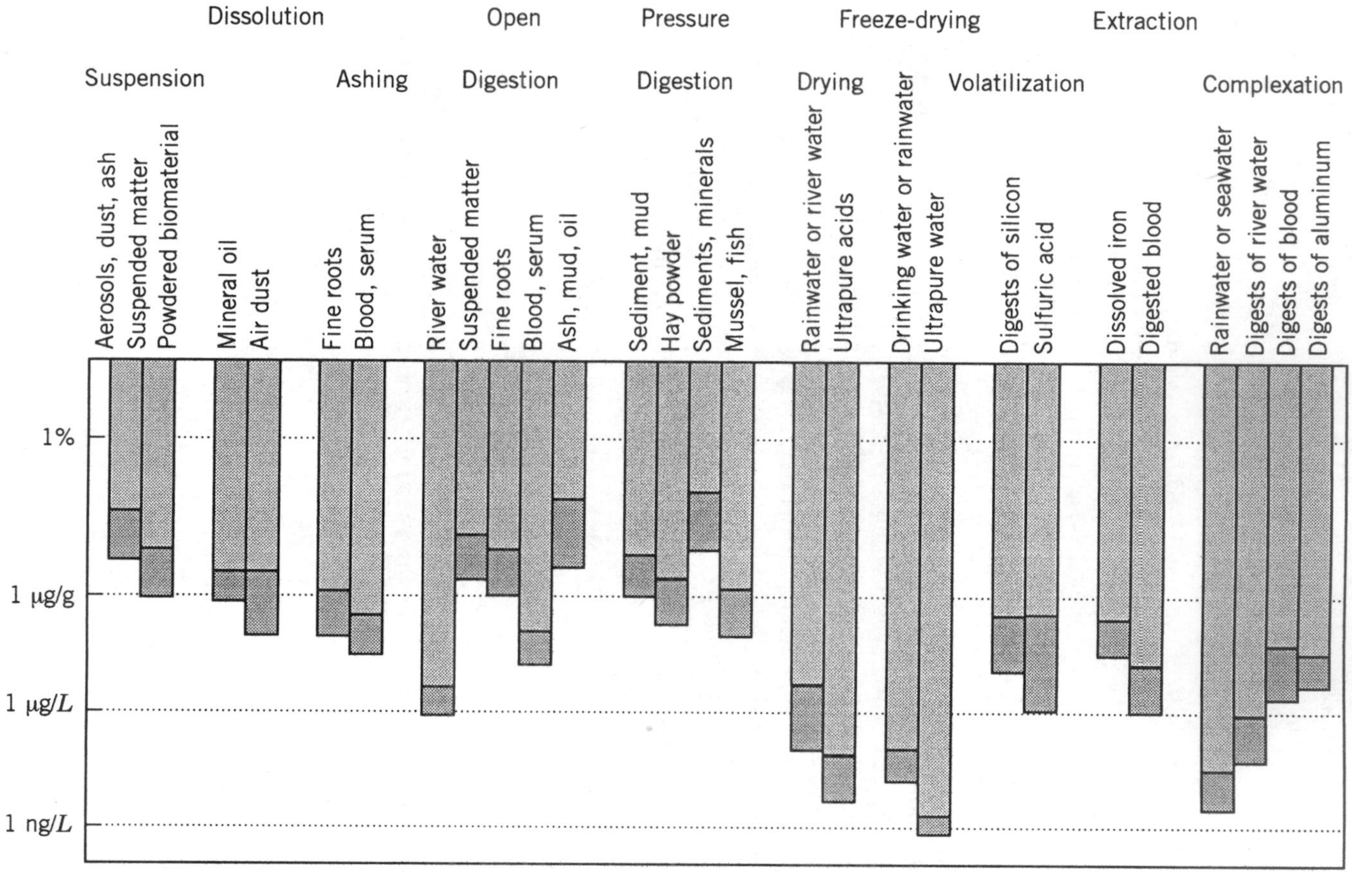

Figure 5-2. Actual detection limits of TXRF measured for real samples after a specific preparation. Because of various analyte elements, the detection limits have a range of uncertainty that spans 1 order of magnitude.

setting of parameters. But up to 6 orders of magnitude can be reached if the tube current and live time are simply changed.

Precision and accuracy are the decisive figures of merit for quantitative analyses. The *precision* can be characterized by the relative standard deviation of repeated quantitative determinations; the *accuracy*, by the mean relative deviation of the actual determinations from the nominal or true value.

According to equation (4-8), the concentration of an analyte is determined by the ratio of net intensities. Consequently, multiplicative errors are compensated and instrumental fluctuations, e.g., of current or voltage, do not influence the results. Only the photon counting is decisive, which is controlled by Poisson statistics. The net intensities of both the analyte peak and the internal standard peak are obtained by subtracting the spectral background from the respective gross intensities. Because of this subtraction, nonspecific additive errors are eliminated that are caused, for example, by different carriers. Such fluctuations do not affect the result, which again is only influenced by the photon noise.

The relative standard deviation s_{rel} of the concentration c_x follows from equation (4-8) and is given by

$$s_{rel}(c_x) = s_{rel}\left(\frac{N_n}{N_{is}}\right) \tag{5-4}$$

where N is the net intensity of either the analyte, x, or the internal standard, is. Due to the counting statistical fluctuations, it follows that

$$s_{rel}(c_x) = \sqrt{\frac{s^2(N_x)}{N_x^2} + \frac{s^2(N_{is})}{N_{is}^2}} \tag{5-5}$$

where $s(N_x)$ and $s(N_{is})$ are the absolute standard deviations of the respective net intensities. These quantities are determined by the square root law:

$$s(N_x) = \sqrt{N_{x+b} + N_{b,x}} \tag{5-6}$$

$$s(N_{is}) = \sqrt{N_{is+b} + N_{b,is}} \tag{5-7}$$

where N_{x+b} and N_{is+b} are gross intensities, and $N_{b,x}$ and $N_{b,is}$ are the background corrections, either of the analyte, x, or the internal standard, is. If we take these relations into account, it follows that

$$s_{rel}(c_x) = \sqrt{\frac{N_{x+b} + N_{b,x}}{(N_{x+b} - N_{b,x})^2} + \frac{N_{is+b} + N_{b,is}}{(N_{is+b} - N_{b,is})^2}} \tag{5-8}$$

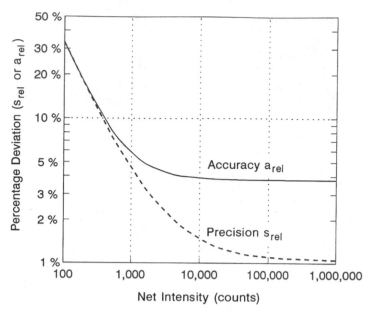

Figure 5-3. Precision and accuracy represented by a percentage deviation, dependent on the net intensity measured for the analyte peak. The net intensity for the internal standard was assumed to be 10,000 counts; the spectral background was estimated to be 500 counts.

This relationship is illustrated in Figure 5-3 for a typical set of intensity values. The background intensities are assumed to be 500 for both the analyte and the internal standard; the net intensity of the internal standard is set to 10,000, and that of the analyte is varied between 100 and 1,000,000. The concentration c_x of the analyte does not appear explicitly but is proportional to the associate net intensity N_x. Determinations significantly above the detection limit can be carried out with a precision of 1–5%.

The deviations between the actual determinations and the nominal values which characterize the accuracy of an analytical method will surpass these values. From equation (4-8), they can be estimated according to

$$a_{rel}(c_x) = \sqrt{s_{rel}^2\left(\frac{N_x}{N_{is}}\right) + a_{rel}^2\left(\frac{S_x}{S_{is}}\right) + a_{rel}^2(c_{is})} \qquad (5\text{-}9)$$

where $s_{rel}(N_x/N_{is})$ is the aforementioned relative standard deviation; $a_{rel}(S_x/S_{is})$ is the relative uncertainty in the determination of the relative sensitivities; and $a_{rel}(c_{is})$, that of the internal standard concentration. These quantities are estimated to be 3% and 2%, respectively, so the relative deviation $a_{rel}(c_x)$ can be calculated. In addition to s_{rel}, this quantity a_{rel} is represented in Figure 5-3. From there, we can see that an accuracy of about 3–6% can be expected for TXRF analyses, significantly above the detection limit.

In the most favorable cases, e.g., for the analysis of pure waters or fine powders, such precision and accuracy can be realized, as will be shown in Section 5.2. Larger systematic errors can occur if a chemical pretreatment of the sample is necessary, e.g., a digestion or matrix separation. Several inter-comparison tests, however, demonstrate a high reliability of TXRF and confirm its competitiveness with the well-established methods of INAA (instrumental neutron activation analysis), ICP-OES, or AAS (e.g., see Michaelis et al. [4, 5] and Burba et al. [6]).

In cases where the results found for a small specimen need to be transferred to a larger sample, that sample has to be homogeneous or must first be homogenized. Only in that case can microanalysis by TXRF give representative results for the total sample. The sampling error for TXRF will be negligible if the sample material is homogeneous below the micrometric scale. This condition can easily be met for liquids or complete solutions but is difficult to fulfill for suspensions, aerosols, or powdered materials. Solid samples generally show coarse-grained inhomogeneities of some $10\,\mu m$ and must be thoroughly ground and/or digested prior to analysis. This requirement is dropped for a true microanalysis where the results must not be transferred to a larger sample. But in any case, the analyte and internal standard should be homogeneously distributed within the small specimen in order to provide reliable quantitative results.

5.1.4. The Great Variety of Suitable Samples

Generally, all liquids and solids will be suitable for analysis if they can be placed onto a sample carrier in a small or even minute amount. Individual particles like grains, fibers, shavings, or filings can simply be put down, powders can be dusted, foils or sections laid down, and layers deposited on a sample carrier. Liquids, solutions, or suspensions can easily be pipetted onto a hydrophobic carrier but have to be dried prior to analysis.

Sample materials of many varieties have been analyzed by TXRF. A diversity of waters [7–13], but also acids [14, 15], oils [16–20], and body fluids [21–27], are just part of the *liquids* that have been investigated. The analyzed *solid* materials are either inorganic and mostly have an industrial or geogenous origin [28–38] or are organic and mostly biogenous [39–44]. Table 5-1 gives an overview of various sample materials. The samples can be analyzed directly, but mostly a suitable preparation step is carried out previously. To get a representative result, the sample should first be homogen-

Table 5-1. Sample Materials Already Analyzed by TXRF

Liquids	Solids (Anorganic)	Solids (Biogeneous)
• *Waters:* drinking water, river water, rain-, sea-, wastewater	• *Suspended matter:* aerosols, dusts, fly ash, organic	• *Plant materials:* algae, hay, leaves, lichen, moss, needles, roots, wood
• *Body fluids:* blood, serum, urine	• *Soils:* mud, sediments, sewage sludge • *Crystals and minerals:* ores, rocks, silica, silicium	• *Foodstuff:* fish, flour, fruits, crab-, mussel meat, mushrooms, nuts, vegetables
• *Pure chemicals:* acids, bases, salts, solvents, water • *Oils and greases:* crude oil, essential oil, fuel oil, fat and grease	• *Metals:* aluminum, iron, steel • *Pigments:* cream, ink, oil paints, powder • *Thin deposits:* contaminations, films, foils, layers, residues	• *Tissue:* hair, kidney, liver, lung, nails

ized. To reach the smallest possible detection limits, the sample matrix should first be separated.

Basically TXRF is a microanalytical method. About 100 μL of a liquid, some 200 μg of an anorganic solid sample, and about 50 μg of an organic biotic sample can be used at the maximum. Only a low level of totally dissolved solids can be used. When a maximum of 1% is exceeded, the detection limits get distinctly worse and are just comparable to those of conventional XRF [45].

TXRF can be applied to micro-, trace-, and surface- or near-surface-layer analyses. Since only a small sample amount is required, TXRF is extremely suitable for microanalyses. These will be necessary if the sample material is valuable or unique or if only a small sample amount is available. Because of the low detection limits, TXRF can also be employed for trace or ultratrace analyses. For that purpose, a larger sample amount of several mL or mg is first homogenized and then one or several small specimens are taken and analyzed. For optically flat samples like wafers, surface and near-surface-layer analyses are performed. A special device for fine-angle setting is needed to record angle-dependent intensity profiles. Furthermore, a fundamental parameter method is needed to evaluate these profiles.

It should be emphasized that the equipment is rather easy to use. The development of a suitable method for resolving a particular problem, especially that of the sample preparation, initially needs the experience of a specialist. But relatively unskilled personnel can operate the instruments, which partly work under computer control, and can use the method for routine analyses. An automated design with a sample changer can facilitate an unattended operation, and even overnight operation is possible. The sample throughput is typically 10 samples per hour for the whole set of elements. Extreme trace analyses take a longer time.

Because of the necessary sampling, TXRF in general is not applied nondestructively. But TXRF is a *nonconsumptive* method because the specimens are not used up. After an analysis, the specimen is still available for further repeat determinations. This possibility may be important, e.g., for the analysis of forensic pieces of evidence and in general for reference analyses.

5.2. ENVIRONMENTAL APPLICATIONS

The analytical features of TXRF have promoted its use in different kinds of applications for several types of matrices. Lots of applications concern samples in the field of environmental studies, where TXRF can be applied to pollution control. In particular, the applicability of TXRF may be illustrated by the following examples.

TXRF is especially suitable for ultratrace analyses of pure waters such as rainwater or drinking water. The low-ng/mL level is directly accessible. For river water and seawater, as well as for wastewaters, a preparation is recommended to separate the suspended matter and to remove the salt content. TXRF is also suitable for the analysis of airborne particulate matter. Collection of air dust can be carried out by filtration or impaction. Furthermore, not only inorganic materials such as ashes, sludges, sediments, and soils but also organic materials such as fruits, cereals, grasses, and vegetables can be analyzed after digestion and preferably after matrix removal.

The analysis of biological matrices can have a nutritional aspect as well as an environmental one. The *nutritional* aspect deals with the assimilation and metabolism of various foodstuffs by the organism. The effects that certain elements and in particular some element species have on health are examined. The *environmental* aspect is concerned with pollution and its monitoring by appropriate plants like moss or lichen [46, 47].

The most usual preparatory technique for plant materials is nitric acid digestion of freeze-dried and pulverized components. If volatile elements have to be determined, a pressure digestion is preferable. Otherwise, an open digestion or even a cold plasma ashing can be carried out. Plant extracts like vegetable oils were analyzed after a simple dilution [46]; aqueous extracts like tea were analyzed directly [48]. Cellular extracts of vegetables, called cytosols, were investigated after gel-permeation chromatography [44]. About 80 fractions were separated and only 0.5 mL of each fraction were directly analyzed by TXRF. Several metallic species could be distinguished, thereby elucidating the nature of metal complexing agents.

5.2.1. Water Samples

Pure waters such as rainwater and drinking water, including tap and mineral water, can be analyzed *nearly* directly. Only a few easy preparatory steps have to be taken, as already shown in Figure 4-12. A volume of about 100 mL is first provided, and an aliquot of 1 or 2 mL is acidified with nitric acid (pH 2). A single-element standard, e.g., Se, Co, or Ga, is added on the μg/mL level with a nitric acid base. The standard is homogeneously mixed and aliquots of 10–100 μL are pipetted onto cleaned hydrophobic sample carriers. The droplets are dried on a hot plate or under infrared light, and the residue of 0.1–10 μg is analyzed within a counting time of 100–1000 s. Quantification is carried out according to equation (4-8).

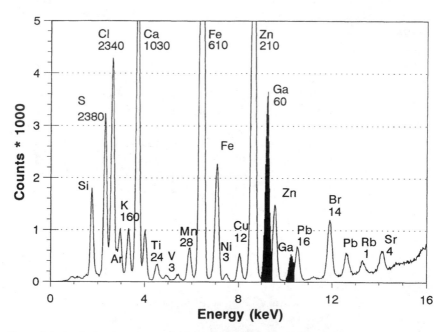

Figure 5-4. Typical TXRF spectrum of a rainwater sample. Gallium was added as the internal standard with a concentration of 60 ng/mL. All values indicated in ng/mL.

Figure 5-4 shows a typical spectrum of rainwater. Detection limits go down to some ng/mL. Table 5-2 shows the quantitative results of TXRF applied to the reference material NIST[3] 1643c "Water" [13, 46]. For these studies, no further preparatory steps were necessary besides the addition of Y as the internal standard and the evaporation of 50 µL on a clean carrier. Mo- and W-excitation were used, with the counting time chosen to be 1000 s. Repeated determinations show a precision of 1–6% (squared mean, 4%). The relative deviations from the certified values reveal an accuracy of 1–12% (squared mean, 6%). For most elements, measured and reference values agree quite well; for only Cr, Sr, and Mo a relative deviation of about 8% is significant.

Table 5-2. TXRF Results for Various Elements Determined in NIST 1643c "Water"[a]

Element	Certified Value (µg/L)	TXRF Result (µg/L; $n = 3$)	Deviation (%)	Significant Distinction
K	2300	2280 ± 80	− 0.9	No
Ca	36800 ± 1400	35300 ± 500	− 4.1	No
V	31.4 ± 2.8	28.5 ± 2.1	− 9.2	No
Cr	19.0 ± 0.6	17.2 ± 0.2	− 9.5	Yes
Mn	35.1 ± 2.2	32.6 ± 1.3	− 7.1	No
Fe	106.9 + 3.0	102.4 + 4.5	− 4.2	No
Co	23.5 ± 0.8	22.3 ± 1.1	− 5.1	No
Ni	60.6 ± 7.3	60.0 ± 0.4	− 1.0	No
Cu	22.3 + 2.8	22.1 + 0.6	− 0.9	No
Zn	73.9 ± 0.9	74.4 ± 0.6	+ 0.7	No
As	82.1 ± 1.2	79.7 ± 1.4	− 2.9	No
Se	12.7 ± 0.7	12.0 ± 0.2	− 5.5	No
Rb	11.4 ± 0.2	11.5 ± 0.4	+ 0.9	No
Sr	263.6 ± 2.6	248.7 ± 3.9	− 5.7	Yes
Y	Internal standard	100.0	—	—
Mo	104.9 ± 1.9	95.7 ± 3.7	− 8.8	Yes
Ag	20.21 ± 0.30	1.90 ± 0.26	14.0	No
Cd	12.2 ± 1.0	11.3 ± 0.6	− 7.4	No
Te	2.7	< 5	—	—
Ba	49.6 ± 3.1	47.6 ± 2.4	− 4.0	No
Tl	7.9	8.6 ± 0.5	+ 8.9	(No)
Pb	35.3 ± 0.9	34.2 ± 1.7	− 3.1	No
Bi	12	13.5 ± 0.8	+ 12.5	(No)

Source: After Reus et al. [13].

[a] The relative deviation of these values from the certified values is given in percent. Both values are called significantly distinct (Yes) or not (No) if their regions of confidence ($x ± s$) do not overlap or do, respectively.

---

[3] NIST stands for the National Institute of Standards and Technology (Gaithersburg, Maryland).

The simple method just described can be improved by freeze-drying of a 10 mL volume and leaching the residue with 1 mL diluted nitric acid. Detection limits are lowered to about 10–20 pg/mL [7].

For river water, estuarine water, and seawater as well as for wastewater, detection limits can be brought down to ng/mL, but some preparation is necessary. If an organic load is significant, a prior pressure filtration is recommended [12]. The filtrate of some 10 mL can be freeze-dried and the residue digested by nitric acid. If a salt matrix is troublesome, the trace elements can be separated by complexation, chromatographic adsorption, and subsequent elution—a method developed by Prange et al. [8, 12, 49]. Complexation of the traces with a NaDBDTC (sodium dibutyldithiocarbamate) solution leads to coprecipitation. Subsequent adsorption of the carbamate traces by a reverse-phase column (e.g., Chromosorb, E. Merck, Darmstadt, Germany) and a final elution of the adsorbed complexes by 2 mL of subboiled chloroform gives an enriched eluate (by a factor of about 50). After the final solution is spiked with an internal standard, an aliquot of 10–100 μL is analyzed as already described. The detection limits for the aqueous solutions including the filtrate are about 0.1 ng/mL.

The suspended matter separated from the filtrate can be collected on Nuclepore filters (Nuclepore Corp., Pleasanton, California), weighed, and digested by concentrated nitric acid or a mixture of nitric and hydrofluoric acid (2:1). The determination by TXRF can be carried out as usual. Detection limits go down to 5 μg/g; precision and accuracy are characterized by a relative standard deviation of about 10% [12].

Recently, some national and international programs have been carried out for pollution control of rain water, river water, and seawater. The River Elbe [12, 50] was tested systematically for contaminants, and appropriate actions were taken for regeneration. Trace contaminants in the Atlantic Ocean were studied at several deep-water stations [51], and heavy metal traces and pollutant transfer were investigated in the North Sea [10, 52]. These field experiments demonstrated the ability of TXRF to automatically handle scores of water samples with a high degree of reliability. The quality of TXRF determinations was proved by different reference samples and by various intercomparison tests of seawater and limnetic or estuarine river water [9, 12, 53]. In competition with such well-established methods as voltammetry, INAA, and ET-AAS, TXRF showed a high level of performance for the certification of reference materials.

For routine analyses of sediments, microwave digestion with nitric acid can also be recommended [35]. Up to 20 elements were determined in sediments of the German shoals ("Wattenmeer"), i.e., for the fine-grained fraction of < 20 μm diameter. The mass fractions ranged from mg/g down to ng/g, covering 5 orders of magnitude.

5.2.2. Airborne Particulates

Dust can be collected from air either by filtration or impaction. In the first method, a certain volume of air is pumped through a filter while the aerosol droplets and particulates are deposited on the filter. If Nuclepore filters are used, the sampled material can be removed by ultrasonic treatment with a nitric acid solution. The mixture of dissolved and particulate matter can be analyzed by TXRF after internal standardization. If membrane filters are used, the loaded filter plus the internal standard can be subjected to pressure digestion with several mL of 65% nitric acid, as described in Section 4.1.2. An aliquot of $10-100\,\mu L$ of the final solution is analyzed by TXRF as usual.

In the second method, air is pumped through fine nozzles and dust particles are deposited onto impaction plates due to their inertia. Different stages with nozzles of decreasing diameter are stacked on top of each other so that particles of different size classes can be collected on different impaction plates. The most common impactors are Berner-, Anderson-, or Battelle-type impactors and have 2–12 stages. Membrane filters are normally used as impaction plates in a Berner impactor. They can be prepared and analyzed as just mentioned. Simple plastic plates are applied in Anderson or Battelle impactors. Consequently, Plexiglas carriers fitted for TXRF devices can be used. In order to preserve the original geometry of the impactor, the carriers are inserted in ring-shaped supports. Figure 5-5 shows a cross section of a two-stage Battelle impactor [54].

Airborne particulates of wet air or aerosols are reliably deposited and stick to the flat carriers [55]. Particles of dry air, however, are bounced or blown off. To avoid these effects, the Plexiglas carriers should first be coated with a suitable grease. Medical petrolatum (e.g., Vaseline petroleum jelly) is recommended for its high purity [56]. An alkaline Ge-standard can be pipetted onto the coated carrier and dried by evaporation. After a blank control, the carriers are inserted into the impactor and used for sampling. The dust laden carriers are directly analyzed by TXRF.

Figure 5-6 shows a spectrum of dust from ambient air collected by a two-stage Battelle impactor [54]. The various elements were found in the nanogram range, with detection limits of <0.1 ng. They are 3 orders of magnitude less than with conventional XRF. Consequently, the collection time can be reduced to 1 h and the sampling volume to $0.5\,m^3$, so that pollution in the course of a day can be observed.

Due to such direct analysis by TXRF, otherwise frequent systematic errors of digestion or dissolution are avoided. However, the high sensitivity of TXRF reveals other systematic errors. In a particular study, air was sampled from a clean bench but nevertheless led to significant contamination of previously clean carriers [54, 56, 57]. These blank values were caused by erosion of the

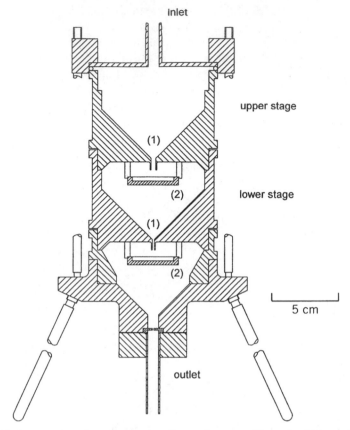

Figure 5-5. Cross section of a two-stage Battelle-type impactor: (1) nozzles; (2) impaction plates made of Plexiglas (After Bayer et al. [54].)

impactor walls made of stainless steel. The error can be avoided by use of a more suitable impactor material, e.g., an antistatic polymer.

Effective monitoring of air pollution presupposes small sampling volumes and short collection times, as is offered by TXRF. This advantage has been used to study the widespread deposition of atmospheric pollutants into forests [58, 59]. Multielement analyses by TXRF were performed with great success.

5.2.3.   Plant Materials

The individual components of plants, e.g., leaves, needles, stalks, blades, rosettes, or roots, are thoroughly cleaned, shredded, freeze-dried, and finally pulverized in a cleaned porcelain mortar with a pestel [43, 44]. Then 100 mg of

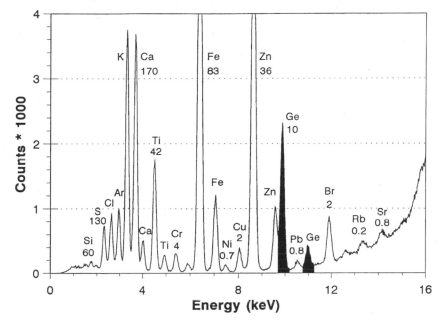

Figure 5-6. TXRF-spectrum of air dust collected from ambient air. A Battelle-type impactor was used for collection of particles with diameters between 2 and 4 μm. Germanium was added as the internal standard of 10 ng. The other specifications are given in ng/m³.

the powdered and possibly sieved plant material is doused with 3 mL HNO_3 (65%, subboiled). The mushy mixture is digested for 3 h either in a small beaker at 110 °C [13,46] or in a PTFE bomb at about 200 °C and 10 MPa [47]. After addition of an internal standard, dilution to 5–20 mL, and cooling, there is either a clear or a slightly opaque solution. The latter indicates that digestion is not totally complete and that a supplementary ashing may be needed. Digestion with sulfuric acid is not suitable for TXRF since the droplets pipetted on the carriers would hardly evaporate.

The foregoing method was applied to the reference material NIST 1573 "Tomato Leaves" [13, 46]. After open digestion, 20 μL of the final solution was pipetted on cleaned quartz glass carriers and dried on a hot plate. Measurements were performed with a commercial instrument at standard settings and a counting time of 1000 s. The results demonstrated in Table 5-3 show a satisfactory precision (2–15%) and a sufficient accuracy (2–30%) for most elements. Only for the elements Ni, Zn, Mo, and Pb is the deviation of certified and measured mass fractions significant. These values, however, were determined at a detection limit of some μg/g.

Table 5-3. TXRF Results Determined for Reference Material NIST 1573 "Tomato Leaves" After Open Digestion[a]

Element	Unit	Reference Value	TXRF Result ($n = 3$)	Deviation (%)	Significant Distinction
P	mg/g	3.37 ± 0.22	3.44 ± 0.28	$+2.1$	No
S	mg/g	6.2 ± 0.4	5.99 ± 0.39	-3.4	No
K	mg/g	44.4 ± 2.4	43.8 ± 2.6	-1.4	No
Ca	mg/g	28.3 ± 2.3	23.5 ± 3.0	-17.0	No
Sc	mg/g	0.00017	$= 2.00$ int. standard	—	—
Ti	μg/g	56 ± 39	79 ± 2	$+41$	No
V	μg/g	1.2 ± 0.2	<4	—	—
Cr	μg/g	4.0 ± 0.5	3.1 ± 0.4	-23	No
Mn	μg/g	224 ± 13	231 ± 10	$+3.1$	No
Fe	μg/g	580 ± 110	$600. \pm 80$	$+3.4$	No
Ni	μg/g	1.3 ± 0.2	3.2 ± 0.4	$+146$	Yes
Cu	μg/g	11 ± 2	12.8 ± 0.6	$+16.4$	No
Zn	μg/g	61 ± 4	70.1 ± 3.2	$+14.9$	Yes
As	μg/g	0.25 ± 0.04	<2	—	—
Se	μg/g	0.054 ± 0.006	<1	—	—
Rb	μg/g	17.3 ± 2.5	18.0 ± 0.9	$+4.0$	No
Sr	μg/g	42 ± 5	43.6 ± 3.3	$+3.8$	No
Zr	μg/g	—	3.1 ± 0.9	—	—
Mo	μg/g	0.53 ± 0.09	1.1 ± 0.4	$+108$	Yes
Cd	μg/g	2.5 ± 0.2	3.3 ± 0.9	$+32$	No
Sb	μg/g	0.036 ± 0.007	<7	—	—
Ba	μg/g	57 ± 9	43 ± 11	-25	No
Ce	μg/g	1.3 ± 0.2	<10	—	—
Pb	μg/g	5.9 ± 0.8	8.6 ± 1.6	$+46$	Yes
U	μg/g	0.059 ± 0.006	<2	—	—

Source: After Reus et al. [13].

[a] Scandium was used as the internal standard; Mo- and W-excitation.

5.2.4.   Vegetable and Essential Oils

The simplest preparation consists of a dilution of oil samples and has been checked for soya oil, peppermint oil, lime oil, and cumin oil [18, 46]. An aliquot of 1 mL is diluted with toluene (1:1), and an internal standard, e.g., organo-Cu or -V in toluene or an oil-based standard (E. Merck, Darmstadt, Germany), is added to a concentration of 10 μg/mL. Then 20 μL of this solution is pipetted onto a quartz glass carrier, which is heated to 100 °C on a hot plate for 5 min to remove the volatile parts of the matrix. The dry residue is analyzed by TXRF within 1000 s. Detection limits are in the range of 3–20 ng/mL.

In order to improve the detection power, a cold plasma ashing is recommended [46]: 1 mL of an oil sample containing the internal standard is bottled in a small quartz beaker and ashed in a low-temperature oxygen plasma (1 h at 300 W and 500 Pa O_2-pressure). The residue is dissolved with 500 μL of half-concentrated nitric acid. Then 50 μL of this solution is applied to TXRF. As compared to the direct method described in the previous paragraph, this method lowers detection limits by about 1 order of magnitude.

But ashing needs about 2 h, and volatile elements like halogens, Hg, Se, Ti, or S are lost and cannot be determined. For these elements, pressure digestion and subsequent plasma ashing are recommended but require about 10 h. For several elements, the detection limits of the aforementioned direct method are fortunately sufficient, in particular with respect to legal regulations for toxic elements.

5.3. MEDICAL AND CLINICAL APPLICATIONS

Trace elements have an important biological function and an impact on all living beings. A depletion of essential elements such as Cr, Mn, Fe, Co, Ni, Cu, Zn, Se, or I will lead to various human deficiency diseases. An accumulation of elements can lead to toxic symptoms or even poisoning and is frequently caused by heavy metals like Cd, Hg, or Pb as environmental contaminants. Accordingly, there is an extensive demand for trace analytical information in the medical and clinical field. Trace analyses are carried out of different materials used as monitors for part or all of the organism. Suitable monitors are body fluids such as blood, serum, plasma, and urine or tissue samples from biopsies of organs, as well as bones, hairs, or nails.

In recent studies, most of these materials have been investigated by TXRF. The method was shown to be especially suitable for trace analyses of whole blood [23, 25], blood serum [21–26], and amniotic fluid [27], and for microanalyses of organ tissue [39, 41, 42], hairs [3, 45], and dental plaque [60]. Simple techniques of sample preparation have been utilized, e.g., microwave digestion with nitric acid (after Prange et al. [23]) or freeze-cutting by a microtome (after von Bohlen et al. [41]) as known from histology. Detection limits down to 20 ng/mL were reported for body fluids and down to 100 ng/g for tissue samples. The reliability of the determinations was checked with the help of certified reference materials and found to be satisfactory. In addition to the advantages of a real multielement determination, the investigators emphasized the ease of quantification and the small sample volume required. The last feature is specially important for biopsies and generally in pediatrics.

5.3.1. Blood and Serum

For both sample materials—whole blood and blood serum—microwave digestion with nitric acid is recommended [23, 46]: 1 mL of the sample is used, and 5 mL of ultrapure nitric acid is added. Digestion is speeded up in a microwave oven for 15 min at 550 W, followed by 20 min at 400 W. After cooling, Co or Ga is added as the internal standard with a mass of 10 μg to reach a concentration of 10 μg/mL sample volume. An aliquot of 10 μL of the final solution is pipetted onto a clean quartz glass carrier and dried by evaporation. TXRF analysis is carried out with a Mo-tube and a counting time of 1000 s. Table 5-4 gives results obtained for the certified reference material NIST 909 "Human Serum". The results for the individual elements generally agree with the certified values except for Ca (relative deviation 10%). The detection limits for heavy metals are about 20–60 ng/mL.

For whole-blood samples, digestion is mandatory. Serum samples, however, may even be analyzed without digestion but after a simple dilution with ultrapure water (1:3). In this case, detection limits are three times higher but may be sufficient for all elements with concentrations above 0.2 μg/mL.

The spectra of whole-blood samples are dominated by the peaks of Fe, K, and Ca impeding the detection of, e.g., Mn, Ni, and Pb. In order to improve the

Table 5-4. TXRF Results for Various Elements Determined in NIST 909 "Human Serum" [a]

Element	Certified Value (μg/mL)	TXRF Result (μg/mL; $n = 3$)	Deviation (%)	Significant Distinction
P	—	153 ± 7	—	—
S	—	1111 ± 49	—	—
K	137.6 ± 4.3	133.4 ± 2.8	− 3.1	No
Ca	120.8 ± 3.5	109.1 ± 1.8	− 9.7	Yes
Cr	0.091 ± 0.006	0.095 ± 0.014	+ 4.4	No
Mn	—	0.099 ± 0.017	—	—
Fe	1.98 ± 0.27	2.14 ± 0.021	+ 8.1	No
Ni	—	0.085 ± 0.021	—	—
Cu	1.10 ± 0.10	1.08 ± 0.02	− 1.8	No
Zn	—	1.21 ± 0.04	—	—
Se	—	0.101 ± 0.006	—	—
Sr	—	0.051 ± 0.006	—	—
Pb	0.02 ± 0.003	0.025 ± 0.012	+ 25	No

Source: After Prange et al. [23].

[a] The relative deviation is given with respect to the certified values. A short test was applied to decide whether the two mean values are significantly distinct (Yes) or not (No).

detection limits for these elements, first Fe is separated and subsequently K and Ca [23]. Initially, Fe is selectively extracted from a 6 N HCl solution with 1 mL MIBK (methyl isobutyl ketone). Then the salt matrix with K and Ca is separated according to the method developed by Prange [8, 49] for seawater analyses; 100 μL of the final solution are applied for the TXRF measurements. The detection limits of this procedure are in the range of 2-5 ng/mL and are sufficient for the determination of Mn, Ni, and Pb in whole-blood samples.

5.3.2. Tissue Samples

Tissue of organs can be analyzed after ashing and/or digestion. It may, however, be preferable to simply cut a tissue sample in thin sections and directly place these sections on TXRF carriers [41, 42]. Besides simplicity, this method offers the advantage of preventing contamination and losses caused by a chemical preparation.

A small piece of tissue with a dimension of < 10 mm and a mass of < 100 mg is frozen at a temperature of − 10 to − 20 °C. The frozen sample is cut by a microtome (Reichert-Jung, Nussloch, Germany) in thin sections each about 15 μm thick. Every section is slid onto the scalpel with the aid of a device called a stretcher. After about 20 s, the section can be placed in the center of a glass carrier by gently touching it with the surface of a cleaned but not siliconized carrier. Figure 5-7 shows a microtome and a thin section of mussel tissue on a quartz glass carrier. Subsequently, the sections are spiked with a droplet of an internal standard, e.g., 10 μL of a Ga standard solution with a 1 μg/mL concentration. The droplet is soaked up by the section and dried by

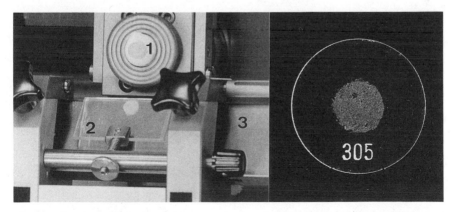

Figure 5-7. Freezing microtome (*left*): (1) tissue sample; (2) stretcher with a thin section; (3) scalpel. A thin section of mussel tissue (*right*) is placed on a quartz glass carrier. (Reproduced from Klockenkämper et al. [42].)

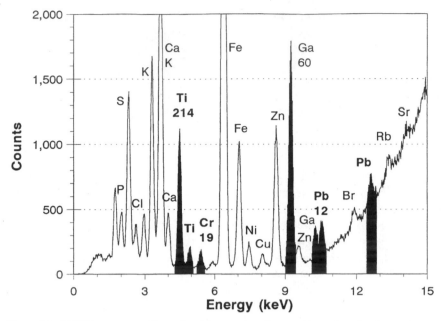

Figure 5-8. TXRF spectrum of lung tissue from a foundryman. Dust deposits show a typical enrichment of Pb, Cr, and Ti with 12, 19, and 214 μg/g, respectively. Ga was used as the internal standard.

evaporation on a hot plate. At the same time, the section shrinks to a thickness of < 10 μm.

The individual sections are analyzed by TXRF within a counting time of 100–1000 s. Afterward, they are dried in a small clean oven at 85 °C for 1 h in order to determine the dry mass. For this purpose, a difference weighing is carried out by means of a microbalance. Every carrier is weighed twice, first *with* the deposited section and then *without* the section, after it is scraped off with a scalpel. The carriers must not be touched by hand, and gloves must be used in order to precisely determine dry masses of only 100 μg or so. The mass fractions are finally determined by equation (4-8).

The freeze-cut technique can be applied to the analysis of tissues, e.g., of kidney, liver, and lung. Figure 5-8 shows a spectrum of lung tissue obtained from a foundry worker. The quantitative results show heavy-metal contamination in the upper μg/g range, which is obviously caused by an occupational exposure to metalliferous dusts.

The reliability of this method was checked by analysis of various reference materials [42]: NIST 1573 "Tomato Leaves"; NIST 1566 "Oyster Tissue";

NIST 1577 "Bovine Liver"; and NIES[4] No. 6 "Mussel Tissue." Since these materials are only available as powders, about 100 mg were first mixed with several μL of ultrapure water and then this powder mash was freeze-cut and analyzed by TXRF. The detection limits were found to be about 1 μg/g. The relative deviations between certified and measured values were about 10% for elements with a mass fraction above 10 μg/g. Any possible Fe-contamination by the iron scalpel was precluded by control tests.

The foregoing method is not restricted to investigations of organ tissues but can also be applied to the analysis of plant and animal foodstuffs, e.g., nuts, mushrooms, and shrimps, and to any compact or liquid biomaterial. The method can be carried out simply and rapidly and is therefore suitable for screening or monitoring in the clinical and environmental fields.

5.4. INDUSTRIAL APPLICATIONS

Because of its simplicity, multielement capability, and detection power, TXRF is highly suitable for the analysis of industrial products. It is the method of choice for quality control of ultrapure reagents needed in the electronics, cosmetics, and pharmaceutical industries. Some high-purity-grade acids, bases, and solvents, including high-purity water, can be analyzed nearly directly down to the level of 0.1 ng/mL [15]. Further applications are concerned with high-purity metals like aluminum [32] or iron [34] and with ultrapure nonmetals like silicon [33, 36] or silica [33]. These solid products can be analyzed in the sub-μg/g region after an appropriate matrix separation.

The TXRF analysis of mineral and synthetic oils is another field of industrial interest, e.g., for the petroleum-refining industry and for engineering. It is especially useful for determination of metal traces, e.g., in light and heavy crude oils [20], in lubricating oils [16, 61], in motor oils [62], and in diesel fuels [63]. Light oils can be applied to the carrier as drops but should first be diluted with chloroform [16] or toluene [62] and then heated. Such a preparation can be highly convenient if detection limits of 1 μg/mL are sufficient. If, however, lower detection limits of some ng/mL are necessary, a cold plasma-ashing with oxygen may be preferable, as already described in Section 5.2.4.

Finally, the TXRF analysis of alloys for major and minor constituents at the percent level is possible after an appropriate digestion, but it is not suitable for production control. In that respect, conventional wavelength dispersive XRF is more convenient and faster and has the highest reproducibility. But TXRF can be used for the control of vaporization and sputter processes by

---

[4]NIES stands for the National Institute for Environmental Studies (Tsukuba, Japan).

which metal or metal oxide layers are prepared. Contaminants can easily be detected [64].

5.4.1. Ultrapure Reagents

Strong acids, bases, or solvents that have a volatile matrix can be analyzed by TXRF nearly directly. Only a few simple preparatory steps are necessary, as already shown in Figure 4-12. A volume of 1 mL is spiked with an internal standard, then a 100 μL volume is pipetted onto a glass carrier in a few steps and evaporated to dryness within 10 min. Even highly concentrated acids like HNO_3, HCl, or HF and bases like NH_3 solutions can be analyzed without dilution. Element impurities can be detected down to or even below 0.1 ng/mL with good accuracy, sufficient for certification [15].

Some useful hints should be noted here: (i) quartz glass carriers must be used for almost all acids, except that pure silicon carriers are needed for hydrofluoric acid; (ii) commercial stock solutions applied for standardization should first be diluted to about 100 ng/mL; (iii) such solutions should be acidified (pH 2) in order to prevent wall-adsorption effects and to avoid hydrolysis.

In order to further improve the detection power, a simple preconcentration step has been recommended [15]. In this procedure, 5 mL of acid is bottled in a Teflon vessel and heated on a hot plate. At the same time, a gentle nitrogen stream removes the acid steam into a water trap. Evaporation down to 1 mL gives detection limits below 20 pg/mL for most elements.

The aforementioned method was applied to some commercial high- purity-grade acids: HNO_3, HCl, and HF. In order to further reduce impurities, an additional purification of acids was carried out afterward by means of a subboiling distillation. The result was checked by TXRF and showed a significant reduction of the impurities. Individual results for the purification of hydrofluoric acid are listed in Table 5-5. The geometric mean of the reduction factors gives a value of 5 as a rough figure of merit.

Reagents like H_2SO_4 or $(NH_4)F$ are not nearly as volatile and need somewhat more complicated preparation. The matrix with a volume of, e.g., 200 μL, can be removed by a low-temperature oxygen plasma, as already proposed for vegetable oils. After about 1 h, the residue can be taken up with some 100 μL of diluted HNO_3 and transferred to a sample carrier for TXRF analysis.

5.4.2. High-Purity Silicon and Silica

High-purity Si and SiO_2 are extremely important products for the semiconductor industry. The determination of impurities on the sub-μg/g level by

Table 5-5. Concentration of Elements in Commercial Suprapure Hydrofluoric Acid Before and After Purification by Subboiling Distillation[a]

Element	Suprapure (ng/mL; $n = 4$)	Subboiled (ng/mL; $n = 6$)	Reduction Factor
P	< 3	< 2	—
S	73.4 ± 9.9	3.0 ± 1.0	24
K	2.33 ± 0.23	1.11 ± 0.19	2
Ca	3.5 ± 1.4	2.3 ± 1.4	1.5
Ti	0.41 ± 0.11	0.35 ± 0.10	1.2
V	< 0.06	< 0.06	—
Cr	0.73 ± 0.02	0.18 ± 0.08	4
Mn	0.11 ± 0.01	< 0.03	⩾ 4
Fe	9.35 ± 0.31	0.40 ± 0.13	23
Co	< 0.04	< 0.02	2
Ni	0.13 ± 0.05	0.09 ± 0.02	1.4
Cu	0.04 ± 0.01	0.04 ± 0.01	1
Zn	0.38 ± 0.03	0.04 ± 0.02	10
Se	< 0.02	< 0.02	—
As	0.23 ± 0.04	< 0.02	⩾ 11
Sr	0.06 ± 0.01	< 0.03	⩾ 2
Zr	< 0.1	—	
Sn	0.3 ± 0.2	—	—
Ba	1.25 ± 0.10	< 0.15	⩾ 8
Pb	4.47 ± 0.05	0.06 ± 0.02	75

Source: After Prange et al. [15].

[a] Rubidium was used as the internal standard.

TXRF is made possible by a dissolution of the materials and a separation of Si as silicon fluoride, which is volatile [33, 36].

Fragments of about 100 mg silicon can be dissolved with a mixture of 0.4 mL HNO_3 (65%, suprapure, subboiled) and 2 mL HF (40%, p.a. or subboiled) in an open PTFE vessel. The chemical reaction can be described by

$$3\,Si + 4\,HNO_3 + 18\,HF \rightleftharpoons 3\,H_2SiF_6 + 4\,NO + 8\,H_2O \qquad (5\text{-}10)$$

For silica, a similar sort of dissolution can be effected by 1 mL HF. The reaction is

$$SiO_2 + 6\,HF \rightleftharpoons H_2SiF_6 + 2\,H_2O \qquad (5\text{-}11)$$

Dissolution occurs in the presence of an excess of HF and by heating the vessel to about 130 °C for nearly 30 min.

After either reaction (5-10) or (5-11), the hexafluorosilicic acid is decomposed according to

$$H_2SiF_6 \rightleftharpoons SiF_4 + 2HF \qquad (5\text{-}12)$$

The volatile fluoride is evaporated by further heating the PTFE vessel at 100 °C for 2 h, thereby reducing the volume of the solution. After evaporation to dryness and cooling, the residue is taken up with 1 mL diluted HNO_3 (10% suprapure) and spiked with an internal standard (Rb or Ga). A portion of 10–50 μL is transferred to a cleaned but not siliconized quartz glass carrier, dried, and analyzed by TXRF.

If impurities of TiO_2 or Al_2O_3 are present, the dissolution will be incomplete. But at any rate, it will lead to a usable fine suspension (grain size < 1 μm). Table 5-6 shows the results found for the certified reference material BCR-CRM[5] 313/1 "High-Purity Silica" [33]. The TXRF values agree quite well with the reference values. Beyond that, about 20 additional elements could be detected. Detection limits were estimated from the blanks, which reached a level of 0.1 μg/g. Only about 1 μg Si of the original matrix of 100 mg was found to be left. Thus, trace elements were enriched by a factor of about 10^5.

5.4.3. High-Purity Aluminum

For the technical evaluation of this material, about 20 trace elements needed to be determined in the range of ng/g to μg/g. To this end, a multistage procedure had to be applied consisting of the digestion of the solid, the adsorption of trace metals as their hexamethylenedithiocarbamates (HMDTC) on a reversed-phase cellulose, and the elution of the collected trace metals [32]. Detection limits of some 10 ng/g could be reached by subsequent TXRF.

For that purpose, 2 g of aluminum are first dissolved in 25 mL HCl (30%, suprapure) by heating at 100 °C for 4 h. The resultant solution is diluted with ultrapure water to a volume of 75 mL. The pH value has to be adjusted to 2.5–3 by addition of 5 mL NaOH in order to prevent hydroxide precipitation. This solution is mixed with 5 mg HMDTC salt (E. Merck, Darmstadt, Germany) dissolved in 100 μL methanol p.a. The final solution is pumped through a small glass column (inner diameter 5 mm) filled with a suspension of 0.1 g acetylated cellulose in 10 mL ultrapure water. A peristaltic pump is used at a rate of 2 mL/min, so that leaching require 40 min.

Afterward, the trace metals loaded on the column are slowly eluted by 0.1 mL methanol, by 1 mL nitric acid (25%), and lastly by 0.9 mL ultrapure

---

[5] BCR stands for the Bureau Communautaire de Référence (Brussels, Belgium).

Table 5-6. Concentration of Impurities Found in the Certified Reference Material BCR-CRM 313/1 "High-Purity Silica" by TXRF After Matrix Removal with HF

Element	Certified Values (μg/g)	TXRF Values (μg/g; $n = 5$)	Deviation (%)	Significant Distinction
Al	190 ± 21	170 ± 15	− 10.5	No
P	—	7.5 ± 2.5	—	—
S	24[a]	20 ± 2	− 17	(No)[b]
K	42 ± 17	49.0 ± 2.0	+ 17	No
Ca	43 ± 7	48.2 ± 0.8	+ 12	No
Ti	100 ± 20	95.3 ± 3.0	− 4.7	No
V	—	0.2 ± 0.1	—	—
Cr	1[a]	0.7 ± 0.09	+ 30	(Yes)[b]
Mn	1 ± 0.2	1.03 ± 0.10	+ 3.0	No
Fe	84 ± 7	83.4 ± 2.5	− 0.7	No
Ni	—	0.43 ± 0.06	—	—
Cu	—	0.58 ± 0.04	—	—
Zn	—	0.34 ± 0.04	—	—
Ga	—	0.02 ± 0.01	—	—
As	—	0.15 ± 0.02	—	—
Rb	—	0.18 ± 0.02	—	—
Sr		1.32 ± 0.07	—	—
Y	—	0.61 ± 0.03	—	—
Zr	15[a]	16.0 ± 1.5	+ 6.7	(No)[b]
Nb	—	0.31 ± 0.07	—	—
Mo		0.07 ± 0.03	—	—
Sn	—	0.10 ± 0.05	—	—
Ba	—	7.9 ± 1.0	—	—
Ce	—	1.6 ± 0.4	—	—
Hf		0.44 ± 0.04	—	—
W	—	0.08 ± 0.04	—	—
Pb	—	0.31 ± 0.04	—	—
Th	—	0.26 ± 0.06	—	—

Source: After Reus [33].

[a] Value not certified, but only recommended.

[b] Parentheses indicate that the test for significant distinction described for Table 5-2 was somewhat changed. Both values are called significantly distinct or not if the recommended value is inside the region of confidence ($x \pm 2s$) of the TXRF value or outside, respectively.

water. The eluate of 2 mL is spiked with an internal standard, e.g., Ga at 200 ng/mL, and aliquots of 20 μL are used for TXRF analysis within a counting time of 200 s.

The precision and accuracy of the foregoing procedure were examined by means of an industrial reference material. Parallel to TXRF, the trace concen-

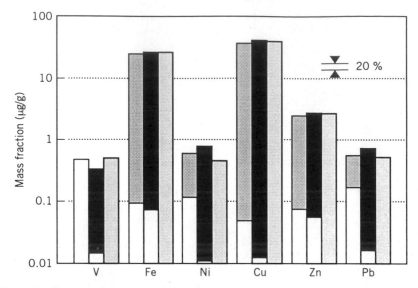

Figure 5-9. Trace metal contents in high-purity aluminum determined by TXRF (solid black) and FAAS (medium gray) in comparison with certified values (light gray). Detection limits are indicated by a white area. The reference material VAW R 14 [65] was chosen for analysis [32]. (VAW = Vereinigte Aluminium Werke, Bonn, Germany).

trate was analyzed by Burba et al. [32] using FAAS and applying the injection technique. Figure 5-9 shows the results to be in a good agreement [65]. Detection limits mainly determined by the blanks of the total procedure are on the order of 10–20 ng/g. The remaining mass of Al is about 80 μg, so that an enrichment factor of 2.5×10^4 was attained.

5.4.4. Wafer Material

Polished wafer disks are the basic material for semiconductor devices. Since they are optically flat and even, they are ideally suited for total reflection of X-rays and consequently are highly suitable for an examination by TXRF. Several commercial instruments have been developed for wafer applications, and today more than 100 are being utilized in the semiconductor industry of the United States and of Japan. Lots of reports have appeared in the last decade, and overviews on the different applications have been given by Hockett [66, 67]. Most studies have been concerned with silicon wafers [68–78], but GaAs wafers have also been evaluated by TXRF (e.g., see Kamakura et al. [79]).

Wafer manufacturing includes a number of different processes. It starts with slicing, etching, polishing, cleaning, and packaging of the raw wafers. Further steps are necessary for IC (integrated circuit) devices, e.g., oxidation, implantation, deposition, and thermal treatment. Recleaning steps may be added in between the other steps. All these steps of the process are capable of some infiltration of contaminants. They may stem from the raw material itself; from gases, chemicals, or tools; from masks or walls of an implanter, furnace, or vaporizer; possibly from a holder; and even from cleaning agents or ambient air. Such contaminations can be the cause of malfunctions of the finished IC devices, especially in VLSI (very-large-scale integration) technology and even more so in ULSI (ultra-large-scale integration) technology. Alkali elements like Na or K can reduce the requisite threshold voltage; transition elements like Cr, Fe, Co, Ni, Cu, and Zn may induce a leakage current; and the actinides U and Th can cause a malfunction by α-emission. To avoid such failure modes, wafers must have extremely low impurities and extremely clean surfaces. Their atomic area density should be $< 10^{10}$ atoms/cm^2.

TXRF is capable of checking the contaminations brought in by the different steps. The aim is to reduce the contaminants to a minimum, to remove the remainder by effective cleaning steps, and finally to improve the manufacturing process. TXRF can be applied to monitor the entire process automatically and to serve as a means of production control. Obviously, TXRF is suited for unpatterned planar wafers rather than for patterned wafers.

The method can be applied directly and nondestructively to determine trace-metal contaminations of $> 10^9$ atoms/cm^2 and even to make a complete map of the total wafer surface. For each point of the surface, only a single measurement at one fixed angle position is necessary. It is even possible to distinguish between particulate and thin-layer-type contaminations. For this purpose, however, at least two measurements at two distinct angle positions are required.

Contaminations down to the level of 10^7 atoms/cm^2 can be determined if they are first collected from the entire surface of a wafer and then concentrated on a small spot. This technique is based on etching of the wafer surface—and consequently requires that the usual goal of nondestructiveness be abandoned in this case.

Finally, TXRF can be employed to characterize thin-layer structures in the near-surface range of 2–500 nm with respect to their composition, thickness, and density. This requires, however, an angle scan of the layered wafer entailing much effort in order to record and interpret angle-dependent intensity profiles.

5.4.4.1. Determination of Wafer Contamination by Direct TXRF

A device is needed for wafer positioning, especially for shifting and fine-angle adjustment, as already demonstrated in Figure 4-13. A W-tube is preferably chosen and the W-$L\beta$- instead of the W-$L\alpha$-peak is selected by a multilayer monochromator in order to excite the transition metals, including Zn, and to avoid the excitation of Ga and As in the case of GaAs wafers. A single measurement is needed for the determination of contaminants at each point of the surface. For this purpose, a fixed glancing angle has to be adjusted that should be about 70% of the critical angle of the wafer material according to equation (4-18). The angle adjustment can be controlled by the fluorescence intensity of the wafer material itself (e.g., silicon), as described in Section 4.4.1. Calibration can be carried out by an external standard (e.g., a Ni-plated wafer), as described in Section 4.4.2.

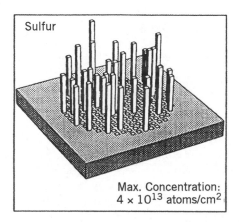

Sulfur

Max. Concentration:
4×10^{13} atoms/cm^2

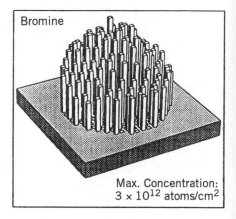

Bromine

Max. Concentration:
3×10^{12} atoms/cm^2

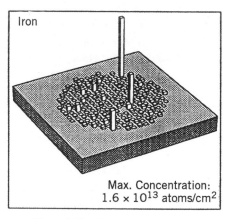

Iron

Max. Concentration:
1.6×10^{13} atoms/cm^2

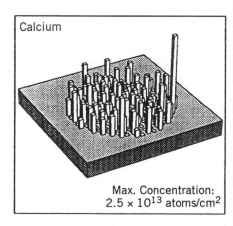

Calcium

Max. Concentration:
2.5×10^{13} atoms/cm^2

Figure 5-10. TXRF results of wafer mapping for the elements S, Br, Fe, and Ca. About 225 different spots with an area of 0.5 cm^2 were investigated on an 8 in. wafer. (Reproduced from Berneike [75].)

Figure 5-10 shows the results of wafer mapping for different elements [75]. The complete mapping with about 225 spots needs about 12 h measuring time. The lateral resolution is determined by the observation area of the detector, which is on the order of 0.5 cm^2. Contaminants are in the range of 10^{11} to some 10^{13} atoms/cm^2. The nonuniform distribution differs for the individual elements. The information it provides may give valuable clues as to the source of these contaminations.

Furthermore, a distinction between two kinds of contamination—the particulate and the thin-layer type—is possible. As described in Section 4.4.2, two angle measurements are sufficient, although a recording of a total intensity profile with 20–30 different positions is preferable. Concrete examples of this approach were first given by Prange and Schwenke [3] and Schwenke and Knoth [80]. As demonstrated in Figure 5-11, the two elements Zn and Br are respectively detected as contaminations of 6.2×10^{11} and 2.7×10^{11} atoms/cm^2. The element Zn is mainly deposited in particulates (85%), whereas Br is deposited nearly exclusively in a thin layer (97%). The analytical error is about 5%.

5.4.4.2. Determination of Wafer Contamination by VPD–TXRF

The detection limits of the direct nondestructive TXRF measurement are on the order of 10^9 atoms/cm^2 for the critical transition metals. They can be improved by more than 2 orders of magnitude, reaching the level of 10^7 atoms/cm^2 if the impurities of the entire surface of the wafer (4–8 in. diameter) are collected and preconcentrated prior to TXRF analysis. The improvement is given by the ratio of the entire area of the wafer and the observation area of the detector.

A common preconcentration technique is vapor-phase decomposition (VPD). Originally developed for AAS measurements, it was later adapted to TXRF determinations [81,82]. The combination of VPD and TXRF is a destructive method but highly sensitive for element traces *on* and *in* the uppermost layers of silicon wafers.

A special VPD reactor, as described by Neumann and Eichinger [82], is schematically illustrated in Figure 5-12. Some 10 mL of a 20% HF solution are poured into a heated Teflon dish and evaporated. The wafer, placed on a Teflon platform and cooled down, is exposed to the HF vapor in the closed reactor. The oxide layer of SiO$_2$, either native and about 3 nm thick or thermally grown and some 10–100 nm thick, is dissolved according to equation (5-11), already given in Section 5.4.2:

$$SiO_2 + 6HF \rightleftharpoons H_2SiF_6 + 2H_2O$$

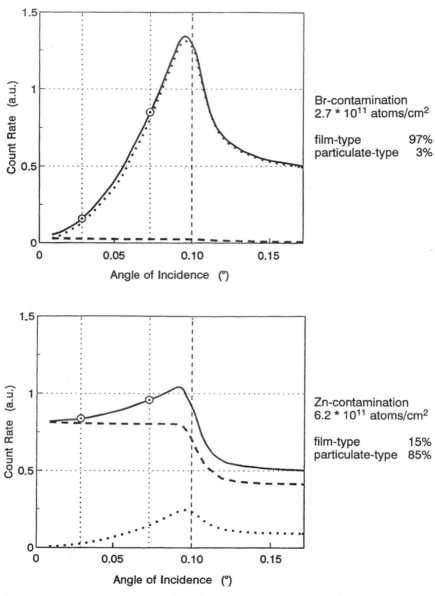

Figure 5-11. Angle-dependent intensity profiles for the elements Br and Zn. The measured profiles (full line) were composed by a particulate curve (dashed) and a thin-layer curve (dotted). Thereby, the total contamination was distributed to a particulate type and a thin-layer type with the percentages shown. (After Prange and Schwenke [3].)

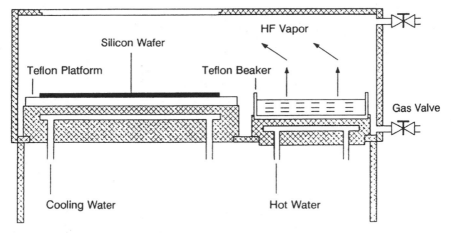

Figure 5-12. Diagram of a VPD reactor suitable for vapor phase decomposition of the oxidized surface of a silicon wafer. (Reproduced after Neumann and Eichinger [82].)

Lots of fine water droplets build a moisture film on the hydrophobic surface within 30 min (up to 3 h). These droplets contain most of the contaminants or impurities previously present on or in the oxide layer of the wafer.

After opening the gas valves and the reactor itself, the wafer is taken out. A drop of 10 or 100 μL deionized water is pipetted onto the surface and moved spirally in order to collect all the fine droplets including the contaminations. Finally, the collecting drop is moved to the center and dried by evaporation. The complete procedure is carried out on a clean bench. Afterward, TXRF is applied for the analysis of the granular dry residue at a fixed-angle adjustment.

Several wafers (up to 10) can be stacked on top of one another and all the droplets may be drained by rocking motions and collected in a glass tub at the bottom [83]. From there, a larger drop can be taken for analysis by TXRF.

Care must be taken to get a residue within an area of < 1 mm diameter and to place it under the center of the detector [84]. For quantification, an external multielement standard can be prepared as a similar residue and measured under the same conditions. Subsequent addition of an internal standard is not recommended since a homogeneous distribution would not be guaranteed. In any case, the residue of metal impurities should have a mass of < 1 μg. Correspondingly, the original contamination of the entire surface must not exceed the level of 10^{13} atoms/cm^2; otherwise negative effects mentioned in Section 4.3.3 will become evident.

A further requirement for accurate results is a high degree of collection efficiency for all the different metal impurities. Most of them show an efficiency of > 70% with the exception of Cu. The Cu atoms are obviously adsorbed on

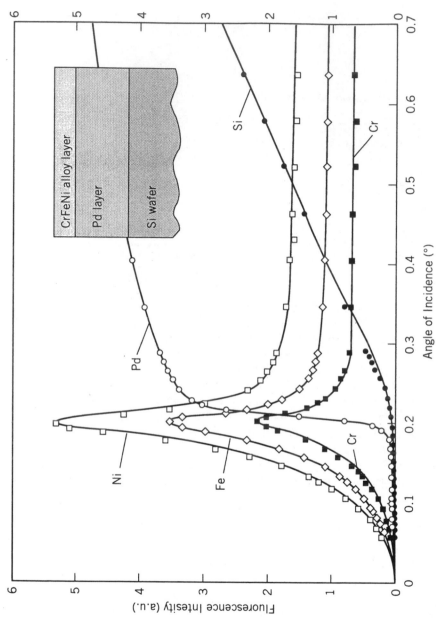

Figure 5-13. Fluorescence intensities of five elements measured on a layered sample by an angle scan. Excitation was performed by the $K\alpha$-peak of a Mo X-ray tube. The solid curves do not represent a regression line of the measuring points but are a best fit of calculations starting with a model demonstrated in the inset on top, right (Taken from Klockenkämper et al. [98].)

the Si surface after the oxide layer is stripped away by the HF vapor. To remove the adsorbed Cu atoms, some hydrogen peroxide H_2O_2 should be added to the deionized water used for collection [85].

5.4.4.3. Thin-Layer Structures

A variety of thin layers deposited on flat surfaces (mainly silicon wafers) have been analyzed by TXRF [86–97]. These single- and double-layer systems have consisted of pure metals, metal alloys, metal oxides, or nitrides. The mass fraction of the individual elements covered the total percentage range, the thickness ranged from about 2 to 500 nm, and the density was between 3 and 12 g/cm^3. Angle-dependent intensity profiles were recorded for these determinations, and the profiles were interpreted qualitatively and calculated quantitatively. The performance already described in Section 4.4.3 can be demonstrated by an example [91, 98].

Figure 5-13 represents intensity profiles of the five elements Cr, Ni, Fe, Pd, and Si measured for a layered sample at glancing angles up to 0.7°. From the similar course of the Cr, Fe, and Ni curves, it can be concluded that there is a first layer consisting of a Cr/Fe/Ni alloy. The course of these curves suggests a thin layer of about 10 nm thickness. The course of the Pd curve refers to a second layer of pure Pd, which in contrast to the first layer is thicker than 100 nm. The Si curve indicates a pure Si substrate.

This first qualitative impression of the curves is a rough estimate, as shown in the inset of Figure 5-13. It serves as a starting point for the quantitative calculations described in Section 4.4.3. The solid curves of Figure 5-13 represent the best fit after several iterations. They are based on a best model presented in Table 5-7 as the final result. It contains the element composition, the thickness, and the density of the two layers and the substrate, thereby making the first rough estimate more sophisticated. The relative standard deviation of the individual data is about 1–2%.

Table 5-7. Final Results of the Fitting Procedure Applied for a Stratified Structure on a Silicon Wafer

Structure	Element Composition (%)					Thickness (nm)	Density (g/cm^3)
	Cr	Fe	Ni	Pd	Si		
Upper layer	46.7	29.5	23.8	—	—	5.9	7.0
Lower layer	—	—	—	100	—	257	11.2
Substrate	—	—	—	—	100	∞	2.3

Source: After Weisbrod et al. [91].

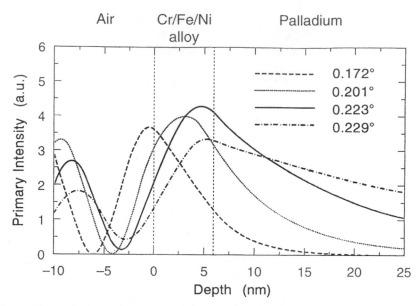

Figure 5-14. Primary intensity above and within a two-layer system as a function of the depth normal to the surface. The intensity was calculated for the layered sample given in Table 5-7 at four different angles of incidence. Excitation by the Mo-$K\alpha$ peak at 17.5 keV was assumed. (Taken from Klockenkämper et al. [98].)

Figure 5-14 demonstrates the standing-wave pattern of the primary beam, which was calculated for the final model and gave the best fit of Figure 5-13. It shows typical oscillations of the intensity dependent on the depth above and within the double layer for different angles of incidence. The standing waves are rather distinct above the surface, especially for angles below the critical angle of the upper layer, i.e., < 0.20°. Within this layer, they are developed only for glancing angles between the two critical angles of both layers, i.e., between 0.20° and some 0.22°. A strong antinode with a more or less fourfold intensity appears. By contrast, a progressive wave appears in the lower Pd layer. It fades out in an exponential manner but deeply penetrates into this layer above the critical angle, i.e., > 0.22°.

5.5. ART-HISTORICAL AND FORENSIC APPLICATIONS

Valuable works of the fine arts have long been investigated by chemical analyses. Such an analytical characterization can obviously be a helpful tool for the purposes of art history and may be useful for answering questions of restoration and conservation. In cases with favorable outcomes, dating of the work of art is made possible and forgeries can be detected.

Various analytical techniques have been applied, and the results obtained are often excellent. But most of the techniques require a small amount of sample material for analysis. Although test portions of a few mg are usually sufficient, such sampling obviously would still damage the work of art. A method that can work nondestructively is conventional XRF. But this method requires a particular bulky apparatus for objects with a dimension of 30 cm or more. In contrast to conventional XRF, the TXRF method needs only μg portions and is therefore virtually nondestructive. A very gentle method of microsampling has been developed that is especially applicable to oil paintings under restoration [99].

In the field of forensic science, microsamples must often serve as pieces of evidence. When an analytical characterization is called for of such small samples, the analyst frequently must make do with mg quantities (microsamples) or even μg quantities (ultramicrosamples). Consequently, TXRF as a microanalytical tool is highly suitable for such forensic questions. In recent years several specific problems have been investigated and appropriate solutions have been developed. These investigations have involved ultramicroanalyses of hair [3,45,46], glass particles [46], and other samples of evidence [100,101], e.g., tape fragments, drug powders, semen traces, and gun-shot residues. Generally, in such cases two similar samples have to be compared and their identical nature must be either confirmed or rejected. For this purpose, minute samples are placed on a glass carrier and normally analyzed by TXRF without any further preparation. The spectral patterns are used as fingerprints of the samples and compared as described in Section 4.2.3.2. It is highly important that there is no consumption at all; the minute samples on the carriers can be preserved as pieces of evidence.

5.5.1. Inorganic Pigments

Oil paints generally are a mixture of pigment powders with an appropriate oil or resin that serves as the bonding agent. A lot of pigments are inorganic, consisting of fine grains of some $0.1-1$ μm. The variety of these pigments is not very large and amounts to only about a hundred different types. Most of them are metal oxides, hydroxides, sulfides, and sulfates or consist of mixed compounds. Each pigment can be characterized by one to five major elements, which can be detected by TXRF, besides the light elements H, C, and O. The detectable elements are usually sufficient for the identification of a particular pigment or paint.

A sampling technique recently proposed for TXRF can be applied to oil paintings either *without* a clear varnish or *with* a varnish that must be removed anyway because of an intended restoration. As already mentioned in Section 4.1.3, a dry cotton-wool bud ("Q-tip") can be used to rub off a minute amount of paint. Different Q-tips are applied to different spots on the painting. This technique is nearly nondestructive since only about 1 µg of paint is rubbed off in each case. The Q-tips are locked into bottle caps made of Plexiglas for safekeeping and transportation. For analysis, they are dabbed onto a sample carrier by a single tip. An amount of less than 100 ng is transmitted, which is enough for TXRF analyses.

The minute sample mass cannot be determined exactly, so an internal standardization is not practicable. Nevertheless, a quantification is made possible by normalization to 100% according to equation (4-9). This method was extended and applied to the analysis of pigment mixtures [99]. The accuracy was checked by a synthetic mixture of three different pigments: titanium white, zinc white, and strontium yellow. The nominal proportion of 1:1:1 was approximately confirmed by the ratio of 0.94:1.04:1.03. The precision was characterized by a relative standard deviation of only 4%.

The foregoing method is fast and convenient and has already been applied to a systematic screening of paintings under restoration, as well as to murals, painted sculptures, and book illustrations [102].

5.5.2. Textile Fibers

A special method has been developed for the analytical characterization of single textile fibers [103]. A variety of 35 different types and models of uncolored textile fibers (e.g., polyester, viscose, or wool) was first analyzed in order to get the respective element patterns. A sample of 500 µg of every kind of fiber was placed on a carrier; then 20 µL of nitric acid with 20 ng of an internal standard (Ga) were added and dried by evaporation. The spectra were recorded directly in order to first determine volatile elements like sulfur. Afterward, a cold oxygen plasma ashing was applied in order to determine the other elements with a higher sensitivity. Mass fractions were found between 10 µg/g and 10 mg/g. The elements K, Ca, Fe, and Zn were regarded as contaminants and not taken into account. But six further elements (P, S, Ti, Mn, Sb, and Ba) were used to establish an appropriate data bank for the total set of 35 chosen fibers.

Subsequently, individual pieces of the set with a length of 2 mm were placed on carriers and analyzed after standard addition. The mass of these small pieces was roughly estimated as 0.7 µg (350 µg/m) in order to determine the mass fractions of the six elements chosen. By application of a 2σ criterion, 65% of the fibers could be identified unambiguously. For the remaining 35%, the number of candidates could be restricted to less than 3 of all the 35 fibers chosen for this investigation.

REFERENCES

1. International Union of Pure and Applied Chemistry (1976). "Nomenclature, Symbols, Units and Their Usage in Spectrochemical Analysis. II. Data Interpretation," *Pure Appl. Chem.* **45**, 99.

2. Reus, U., Freitag, K., Haase, A., and Alexandre, J.F. (1989). *Spectra 2000* **143**, 42.

3. Prange, A., and Schwenke, H. (1992). *Adv. X-Ray Anal.* **35B**, 899.

4. Michaelis, W., Fanger, H.U., Niedergesäss, R., and Schwenke, H. (1985). In *Instrumentelle Multielementanalyse* (B. Sansoni, ed.), pp. 693–710. VCH Verlagsgemeinschaft, Weinheim, Germany.

5. Michaelis, W. (1986). *Fresenius' Z. Anal. Chem.* **324**, 662.

6. Burba, P., Willmer, P.G., and Klockenkämper, R. (1988). *Vom Wasser* **71**, 179.

7. Stössel, R.-P., and Prange, A. (1985). *Anal. Chem.* **57**, 2880.

8. Prange, A., Knöchel, A., and Michaelis, W. (1985). *Anal. Chim. Acta* **172**, 79.

9. Prange, A., Knoth, J., Stössel, R.-P., Böddeker, H., and Kramer, K. (1987). *Anal. Chim. Acta* **195**, 275.

10. Freimann, P., and Schmidt, D. (1988). *Spectrochim. Acta* **44B**, 505.

11. Egorov, A.I., Kabina, L.P., Kondurov, I.A., Korotkikh, E.M., Martynov, V.V., Shchebetov, A.F., and Sushkov, P.A. (1992). *Adv. X-Ray Anal.* **35B**, 959.

12. Prange, A., Böddeker, H., and Kramer, K. (1993). *Spectrochim. Acta* **48B**, 207.

13. Reus, U., Markert, B., Hoffmeister, C., Spott, D., and Guhr, H. (1993). *Fresenius' J. Anal. Chem.* **347**, 430.

14. Reus, U., Freitag, K., and Fleischhauer, J. (1989). *Fresenius' Z. Anal. Chem.* **334**, 674.

15. Prange, A., Kramer, K., and Reus, U. (1991). *Spectrochim. Acta* **46B**, 1385.

16. Bilbrey, D.B., Leland, D.J., Leyden, D.E., Wobrauschek, P., and Aiginger, H. (1987). *X-Ray Spectrom.* **16**, 161.

17. Freitag, K., Reus, U., and Fleischhauer, J. (1989). *Fresenius' Z. Anal. Chem.* **334**, 675.

18. Reus, U. (1991). *Spectrochim. Acta* **46B**, 1403.

19. Schirrmacher, M., Freimann, P., Schmidt, D., and Dahlmann, G. (1993). *Spectrochim. Acta* **48B**, 199.

20. Ojeda, N., Greaves, E.D., Alvarado, J., and Sajo-Bohns, L. (1993). *Spectrochim. Acta* **48B**, 247.

21. Knoth, J., Schwenke, H., Marten, R., and Glauer, J. (1977). *J. Clin. Chem. Clin. Biochem.* **15**, 557.

22. Yap, C.T. (1988). *Appl. Spectrosc.* **42**, 1250.

23. Prange, A., Böddeker, H., and Michaelis, W. (1989). *Fresenius' Z. Anal. Chem.* **335**, 914.

24. Knöchel, A., Bethel, U., and Hamm, V. (1989). *Fresenius' Z. Anal. Chem.* **334**, 673.

25. Ayala, R.E., Alvarez, E.M., and Wobrauschek, P. (1991). *Spectrochim. Acta* **46B**, 1429.

26. Dogan, P., Dogan, M., and Klockenkämper, R. (1993). *Clin. Chem.* (Washington, D.C.) **39**, 1037.

27. Greaves, E.D., Meitín, J., Sajo-Bohns, L., Castelli, C., Liendo, J., and Borgerg, C. (1995). *Adv. X-Ray Chem. Anal. Jpn.* **26s**, 47.

28. Ketelsen, P., and Knöchel, A. (1984). *Fresenius' Z. Anal. Chem.* **317**, 333.

29. Ketelsen, P., and Knöchel, A. (1985). *Staub–Reinhalt. Luft* **45**, 175.

30. Gerwinski, W., Goetz, D., Koelling, S., and Kunze, J. (1987). *Fresenius' Z. Anal. Chem.* **327**, 293.

31. Gerwinski, W., and Goetz, D. (1987). *Fresenius' Z. Anal. Chem.* **327**, 690.

32. Burba, P., Willmer, P.G., Becker, M., and Klockenkämper, R. (1989). *Spectrochim. Acta* **44B**, 525.

33. Reus, U. (1989). *Spectrochim. Acta* **44B**, 533.

34. Chen, J.S., Berndt, H., Klockenkämper, R., and Tölg, G. (1990). *Fresenius' J. Anal. Chem.* **338**, 891.

35. Koopmann, C., and Prange, A. (1991). *Spectrochim. Acta* **46B**, 1395.

36. Klockenkämper, R., Becker, M., and Bubert, H. (1991). *Spectrochim. Acta* **46B**, 1379.

37. Battiston, G.A., Gerbasi, R., Degetto, S., and Sbrignadello, G. (1993). *Spectrochim. Acta* **48B**, 217.

38. Freiburg, C., Krumpen, W., and Troppenz, U. (1993). Spectrochim. Acta **48B**, 263.

39. von Bohlen, A., Klockenkämper, R., Otto, H., Tölg, G., and Wiecken, B. (1987). *Int. Arch. Occup. Environ. Health* **59**, 403.

40. Eller, R., and Weber, G. (1987). *Fresenius' Z. Anal. Chem.* **328**, 492.

41. von Bohlen, A., Klockenkämper, R., Tölg, G., and Wiecken, B. (1988). *Fresenius' Z. Anal. Chem.* **331**, 454.

42. Klockenkämper, R., von Bohlen, A., and Wiecken, B. (1989). *Spectrochim. Acta* **44B**, 511.

43. Günther, K., and von Bohlen, A. (1990). *Z. Lebensm.-Unters.-Forsch.* **190**, 331.

44. Günther, K., and von Bohlen, A. (1991). *Spectrochim. Acta* **46B**, 1413.

45. Prange, A. (1989). *Spectrochim. Acta* **44B**, 437.

46. Reus, U., and Prange, A. (1993). *Application Notes.* Atomika Instruments GmbH, Oberschleissheim, Germany.

47. Schmeling, M., Alt, F., Klockenkämper, R., and Klockow, D. (1996). *Fresenius' J. Anal. Chem.* (to be published).

48. Xie, M.Y., von Bohlen, A., Günther, K., Ma, Y.H., and Klockenkämper, R. (1996). To be published.

49. Prange, A. (1983). Ph.D. thesis, University of Hamburg.

50. Prange, A., von Tümpling, W., Jr., and Niedergesäss, R. (1996). To be published.

51. Schmidt, D., Gerwinski, W., and Radke, I. (1993). *Spectrochim. Acta* **48B**, 171.

52. Haarich, M., Schmidt, D., Freimann, P., and Jakobsen, A. (1993). *Spectrochim. Acta* **48B**, 183.

53. Freimann, P., Schmidt, D., and Neubauer-Ziebarth, A. (1993). *Spectrochim. Acta* **48B**, 193.

54. Bayer, H., von Bohlen, A., Klockenkämper, R., and Klockow, D. (1995). *Mikrochim. Acta* **119**, 167.

55. Schneider, B. (1989). *Spectrochim. Acta* **44B**, 519.

56. Salvà, A., von Bohlen, A., Klockenkämper, R., and Klockow, D. (1993). *Quim. Anal.* **12**, 57.

57 Klockenkämper, R., Bayer, H., and von Bohlen, A. (1995). *Adv. X-Ray Chem. Anal. Jpn.* **26s**, 41.

58 Michaelis, W., Schönburg, M., and Stössel, R.-P. (1989). In *Mechanisms and Effects of Pollutant-Transfer into Forests*, p. 3. H.W. Georgii, ed. Kluwer Academic Publishers, Dordrecht, The Netherlands.

59. Pepelnik, R., Erbslöh, B., Michaelis, W., and Prange, A. (1993). *Spectrochim. Acta* **48B**, 223.

60. von Bohlen, A., Rechmann, P., Tourmann, J.L., and Klockenkämper, R. (1994). *J. Trace Elem. Electrolytes Health Dis.* **8**, 37.

61. Hahn, J.U., and Jaschke, M. (1993). *Staub-Reinhalt. Luft* **53**, 109.

62. Freitag, K., Reus, U., and Fleischhauer, J. (1989). *Spectrochim. Acta* **44B**, 499.

63. Yap, C.T., Ayala, R.E., and Wobrauschek, P. (1988). *X-Ray Spectrom.* **17**, 171.

64. Hoffmann, P., Lieser, K.H., Hein, M., and Flakowoki, M. (1989). *Spectrochim. Acta* **44B**, 471.

65. Kudermann, G. (1988). *Fresenius' Z. Anal. Chem.* **331**, 697.

66. Hockett, R.S. (1994). *Adv. X-Ray Anal.* **37**, 565.

67. Hockett, R.S. (1995). *Adv. X-Ray Chem. Anal. Jpn.* **26s**, 79.

68. Iida, A., Sakurai, K., Yoshinaga, A., and Gohshi, Y. (1986). *Nucl. Instrum. Methods* **A246**, 736.

69. Iida, A., Sakurai, K., and Gohshi, Y. (1988). *Adv. X-Ray Anal.* **31**, 487.

70. Hockett, R.S., Baumann, S.M., and Schemmel, E. (1988). *Proc.–Electrochem. Soc.* **88-20**, 113.

71. Hockett, R.S., and Katz, W. (1989). *J. Electrochem. Soc.* **136**, 3481.

72. Penka, V., and Hub, W. (1989). *Spectrochim. Acta* **44B**, 483.

73. Eichinger, P., Rath, J., and Schwenke, H. (1989). In *Semiconductor Fabrication: Technology and Metrology* (D.C. Gupta, ed.), ASTM 990, p. 305. American Society for Testing and Materials, Philadelphia.

74. Nishihagi, K., Yamashita, N., Fujino, N., Taniguchi, K., and Ikeda, S. (1991). *Adv. X-Ray Chem. Anal. Jpn.* **22**, 121.

75. Berneike, W. (1993). *Spectrochim. Acta* **48B**, 269.

76. Torcheux, L., Degraeve, B., Mayeux, A., and Delamar, M. (1994). *Surf. Interface Anal. SIA* **21**, 192.

77. Gambino, V., Moccia, G., Girolami, E., and Alfonsetti, R. (1995). *Adv. X-Ray Chem. Anal. Jpn.* **26s**, 35.

78. Mori, Y., Shimanoe, K., and Sakon, T. (1995). *Adv. X-ray Chem. Anal. Jpn.* **26s**, 69.

79. Kamakura, T., Sugamoto, H., Tsuchiya, N., and Matsushita, Y. (1995). *Adv. X-Ray Chem. Anal. Jpn.* **26s**, 169.

80. Schwenke, H., and Knoth, J. (1995). *Part. Surf.* [*Proc. Symp.*], *Meet.*, *1992*, pp. 311–323.

81. Huber, A., Rath, H.J., Eichinger, P., Bauer, Th., Kotz, L., and Staudigl, R. (1988). *Proc.–Electrochem. Soc.* **88-20**, 109.

82. Neumann, C., and Eichinger, P. (1991). *Spectrochim. Acta* **46B**, 1369.

83. Arai, T. (1994). Lecture on Wafer Surface Analysis, Rigaku, Osaka, Japan.

84. Yakushiji, K., Ohkawa, S., and Yoshinaga, A. (1993). *Adv. X-Ray Chem. Anal. Jpn.* **24**, 87.

85. Shimono, T., and Tsuji, M. (1989). *Proc. 1st Workshop ULSI Ultra Clean Technol.*, Tokyo, pp. 49–72.

86. Schwenke, H., Knoth, J., and Weisbrod, U. (1990). *X-Ray Spectrom.* **20**, 277.

87. Lengeler, B. (1990). *Adv. Mater.* **2**, 123.

88. Gutschke, R. (1991). *Diploma thesis*, University of Hamburg.

89. de Boer, D.K.G. (1991). *Phys. Rev.* **B44**, 498.

90. Weisbrod, U., Gutschke, R., Knoth, J., and Schwenke, H. (1991). *Fresenius' J. Anal. Chem.* **341**, 83.

91. Weisbrod, U., Gutschke, R., Knoth, J., and Schwenke, H. (1991). *Appl. Phys.* **A53**, 449.

92. de Boer, D.K.G., and van den Hoogenhof, W.W. (1991). *Adv. X-Ray Anal.* **34**, 35.

93. de Boer, D.K.G., and van den Hoogenhof, W.W. (1991). *Spectrochim. Acta* **46B**, 1323.

94. Iida, A. (1992). *Adv. X-Ray Anal.* **35**, 795.

95. Schwenke, H., Gutschke, R., and Knoth, J. (1992). *Adv. X-Ray Anal.* **35B**, 941.

96. Huang, T.C., and Lee, W.Y. (1995). *Adv. X-Ray Chem. Anal. Jpn.* **26s**, 129.

97. de Boer, D.K.G., Leenaers, A.J.G., and van den Hoogenhof, W.W. (1995). *X-Ray Spectrom.* **24**, 91.

98. Klockenkämper, R., Knoth, J., Prange, A., and Schwenke, H. (1992). *Anal. Chem.* **64**, 1115A.

99. Klockenkämper, R., von Bohlen, A., Moens, L., and Devos, W. (1993). *Spectrochim. Acta* **48B**, 239.

100. Nomura, S., Ninomiya, T., and Taniguchi, K. (1988). *Adv. X-Ray Chem. Anal. Jpn.* **19**, 217.

101. Ninomiya, T., Nomura, S., Taniguchi, K., and Ikeda, S. (1995). *Adv. X-Ray Chem. Anal. Jpn.* **26s**, 9.

102. Moens, L., Devos, W., Klockenkämper, R., and von Bohlen, A. (1995). *J. Trace Microprobe Tech.* **13**, 119.

103. Prange, A., Reus, U., Böddeker, H., Fischer, R., and Adolf, F.-P. (1995). *Adv. X-Ray Chem. Anal. Jpn.* **26s**, 1.

CHAPTER

6

EVALUATION AND PROSPECTS

In the preceding chapters, TXRF was represented as a novel variant of energy-dispersive X-ray fluorescence. It was described with respect to its principles, its special instrumentation, and its performance and applications. Several journal reviews and contributions in books have given similar brief or partial reports on TXRF [1 9]. In this concluding chapter, TXRF will be evaluated regarding its efficiency. Its utility for micro- and trace analyses as well as for surface and thin-layer analyses will be emphasized. Its advantages and limitations will be pointed out. The competitiveness of TXRF with other efficient and well-established methods of atomic spectroscopy will be scrutinized. Finally, present trends and future prospects will be considered.

6.1. UTILITY AND COMPETITIVENESS OF TXRF

TXRF is a universal and economical method of multielement analysis. It is a microanalytical tool for small samples or minute specimens that have to be placed on flat totally reflecting carriers. In addition, TXRF is being effectively applied to element trace analyses. Aqueous solutions, high-purity acids, and body fluids are analyzed down to the pg/mL level. Only a few droplets need to be pipetted on carriers and dried by simple evaporation. If a more complex sample is to be analyzed, a decomposition and separation of the matrix becomes necessary. Simple quantification is made possible, since matrix effects do not appear because of the small sample amounts used for analysis. Furthermore, contaminations on flat surfaces can be determined and stratified near-surface layers can be characterized by a depth profile. The method is nondestructive and especially suitable for thin layers of nanometer thickness deposited on wafers. Depth profiles of biogenous materials can be recorded after sectioning of the material with a freezing microtome.

With respect to these capabilities, TXRF has far surpassed the conventional XRF. Indeed, TXRF has attained a leading position in atomic spectroscopy. The outstanding features compete very well with those of instrumental neutron activation analysis (INAA) and inductively coupled plasma–mass spectrometry (ICS-MS). For several applications, TXRF, even has advantages over these methods. In the field of surface and thin-layer analyses, TXRF is

215

able to compete with Rutherford backscattering (RBS) or secondary ion mass spectrometry (SIMS). Moreover, it is unrivaled in the contamination control of wafers and therefore is a widespread analytical tool in the semiconductor industry.

6.1.1. Advantages and Limitations

Several features of TXRF are especially worthy of note. Most of them are advantageous, although some have their limitations. Some features are even drawbacks and are accepted as the price that has to be paid for the benefits. In Table 6-1, the benefits and drawbacks of TXRF are compared.

TXRF's unique capability for micro- or even ultramicroanalyses of small sample volumes should first be mentioned. No additional device is needed for that purpose. On the other hand, an entirely nondestructive analysis is impossible because of the need for minute specimens, although TXRF is a nonconsumptive method like all X-ray techniques. A great variety of samples can be analyzed and a wide scope of applications can be outlined. Solutions, however, must be evaporated prior to analysis. For that reason, nonvolatile liquids are excluded from *direct* TXRF analysis.

Next, TXRF's usefulness for simultaneous multielement determinations should be emphasized. Low detection limits below 10 pg can be reached for about 70 elements, with the exception of low-Z elements. But detection limits

Table 6-1. Benefits and Drawbacks of TXRF Applied to Element Analyses

Benefits	Drawbacks or Limitations
• Unique microanalytical capability	• Impossibility of totally nondestructive analysis
• Great variety of samples and applications	• Limitation for nonvolatile liquids
• Simultaneous multielement determination	
• Low detection limits	• Exception of low-Z elements
	• Limitation by high matrix contents
• Simple quantification by internal standardization	
• No matrix or memory effects	
• Wide dynamic range	
• Nondestructive surface and thin-layer analysis	• Restriction to flat or polished samples
• Simple automated operations	
• Low running costs and maintenance	

will worsen if high matrix contents are present and cannot be separated. Also, detection will be impeded if a neighbor in the periodic table is present at a higher concentration (> 100-fold).

TXRF generally provides a simple means of quantification via an internal one-element standard. No labor-intensive calibration is required. Because of the small sample volume needed for analysis, no matrix effects occur; and as a clean separate carrier is used for each sample, no memory effects occur. Consequently, quantification is easy and reliable and can be performed in a large dynamic range of 5 orders of magnitude.

A new field of TXRF applications has been opened up by the technique that examines a sample under variation of the angle of incidence. It enables nondestructive surface and thin-layer analyses. The technique, however, is restricted to optically flat or polished samples like wafers and glass disks, coated or uncoated. Samples of a certain roughness ($> 0.1\ \mu$m) blur the effect of total reflection used by this technique and so are excluded from investigations.

Finally, TXRF's simplicity of operation should be mentioned, supported by automated sample changing, measurement, and evaluation. The running costs are quite low (e.g., $ 10 per 10 L liquid nitrogen per week), and the maintenance of the instrumentation is rather easy. An X-ray tube that needs to be replaced only after some years costs about $ 4000; a broken window for the detector can be replaced for $ 1000.

Some prerequisites have to be met in order to reap the advantages of TXRF mentioned here. They are compiled in Table 6-2. The capability for microanalysis requires skilful handling of microgram masses or microliter volumes. A great many samples, especially solid samples, can only be analyzed if decomposition is carried out prior to analysis. Low detection limits necessitate matrix separation, and a clean-bench working methodology is a matter of course in ultratrace analysis. A reliable quantification gives representative results only when the total sample is carefully homogenized prior to sampling.

Table 6-2. Advantages of TXRF and Necessary Prerequisites

Advantages	Prerequisites
• Capability for microanalyses	• Handling of micrograms or microliters
• Large variety of samples	• Decomposition of samples
• Low detection limits	• Matrix separation
	• Clean-bench working methods
• Reliable quantification	• Careful homogenization
• Surface and thin-layer analysis	• Exact fine angle control
• Simple unattended operation	• Stable equipment and automation

The possibility of surface and thin-layer analyses presupposes an additional instrumental device for fine-angle control. Finally, a simple and unattended operation is only possible if a stable and compact piece of equipment is employed that is automated and under computer control.

6.1.2. Comparison Between TXRF and Competitors

A great variety of instrumental methods can be applied to element analysis in different fields of investigation. Usually, many elements have to be determined with different atomic numbers, with different concentrations down to extreme traces, and in quite different matrices. Consequently, methods are required which are capable of a multielement detection, have a high detection power, and are applicable to a multitude of sample materials. Moreover, the methods should give accurate results and work economically. With regard to all these features, TXRF plays an important role in atomic spectroscopy [10].

The group of classical methods including gravimetry, titrimetry, voltammetry, and chromatography cannot compete in any way with one of the universal and effective methods of modern spectroscopy like TXRF. Most of the nonspectroscopic methods are only suitable for single-element detection and have only mediocre detection limits—at μg or ng levels. But the simplicity of their equipment and the accuracy of their results ensure them firm places in the analytical laboratory.

Atomic absorption spectrometry (AAS) is still the most common method of atomic spectroscopy. But only the technique using graphite furnaces has low detection limits (at pg levels). This electrothermal AAS (or ET-AAS) is a micromethod like TXRF, but usually only single-element detections can be carried out with it. Furthermore, the standard-addition method used for calibration is rather laborious. A simultaneous multielement detection has been envisaged for laser AAS but not yet worked out. At present, only an oligoelement technique has been developed, using four hollow-cathode lamps and an Echelle spectrometer.

A very common excitation source for optical emission spectrometry is the inductively coupled plasma (ICP-OES). It is a multielement method for macrosamples, applicable to solutions of several mL. But the detection limits are at the ng level (1–100 ng), so that ICP-OES cannot compete with more effective methods like TXRF. The variant using a microwave-induced plasma (MIP-OES) seems to be superior but is still undergoing development and not yet widespread.

Several techniques of laser spectroscopy have recently been developed, based on the absorption, fluorescence, or ionization of a cloud of atomic vapor by means of a strong laser. Some of these techniques permit determinations at fg levels and are very promising for trace analysis. But any comparison would be premature. At present, the high cost of laser systems applicable to the UV spectral region hinder further progress and spread of these approaches.

There are, however, two powerful multielement methods of trace analysis that are highly developed and well established and consequently should be regarded as strong competitors: instrumental neutron activation analysis (INAA), and inductively coupled plasma–mass spectrometry (ICP-MS). Some comparisons of these methods with TXRF have already been made that show alternate priorities, depending on the matrices to be analyzed and the elements to be detected. Intercomparison tests with respect to a variety of environmental samples were first performed by Michaelis [11]. A multielement characterization of tobacco smoke condensates was carried out by Krivan et al. [12], and comparative analyses of spinach, cabbage, and domestic sludge were performed by Pepelnik et al. [13].

Results found for several elements in the standard reference material NIST 1570 "Spinach" are demonstrated in Figure 6-1. The element mass fractions range from about 36 mg/g for K down to 30 ng/g for Hg. The results of the three different methods agree quite well, and most of them are also in agreement with the certified values. The deviations are on the order of 2–30%, generally < 10%, and prove the fairly good accuracy of all three methods.

The detection power of the three methods and additionally of ET-AAS is illustrated in Figure 6-2 [10, 14, 15]. It represents relative detection limits of the different methods applied to trace analysis of aqueous solutions. The figure confirms ICP-MS to have the higher detection power for most elements. But a more detailed evaluation of the competitive methods should consider additional characteristics. Moreover, the analytical problems to be solved have to be taken into account in order to make a careful choice of the most suitable method. Some characteristics, advantages, and disadvantages that are of help in such an evaluation are listed in Table 6-3.

ICP-MS and INAA can be considered as macro- rather than micro-methods, in contrast to TXRF. As demonstrated by Lieser et al. [16], TXRF can have distinct advantages over INAA if extremely small volumes ($10 \mu L$) are to be analyzed on traces. On the other hand, INAA is a nondestructive technique in principle. It is suited for direct analysis of *solids* without any sample preparation. In most reactors, however, it is forbidden to irradiate *solutions* as a safety precaution but only on principle; it is actually possible to irradiate solutions.

In contrast to INAA, TXRF and ICP-MS are rather well suited for analysis of liquids or solutions. Solids first need to be digested or at least dissolved as a suspension. As a result, losses and contaminations can occur. Furthermore, a limitation of the dissolved portion, i.e., a low salt concentration, has to be observed for both ICP-MS and TXRF. Highly concentrated acids or bases

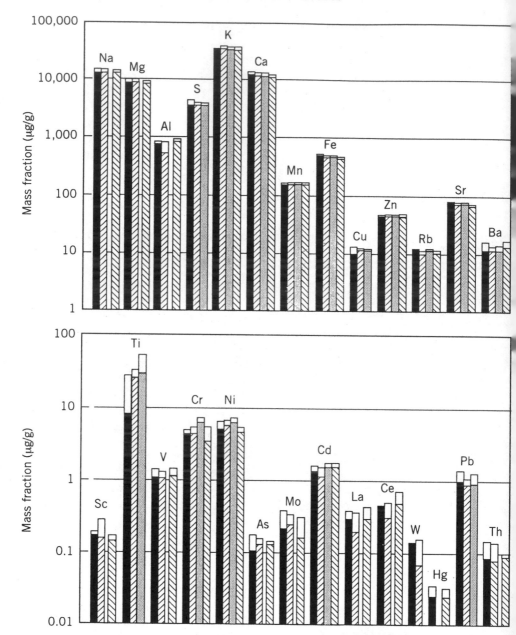

Figure 6-1. Element mass fractions found in NIST 1570 "Spinach" by ICP-MS (left oblique lines), TXRF (mottled gray), and INAA (right oblique lines) in comparison to the certified values (solid black). The bars represent mean values ± standard deviations at $n = 6$ determinations. (After Pepelnik et al. [13].)

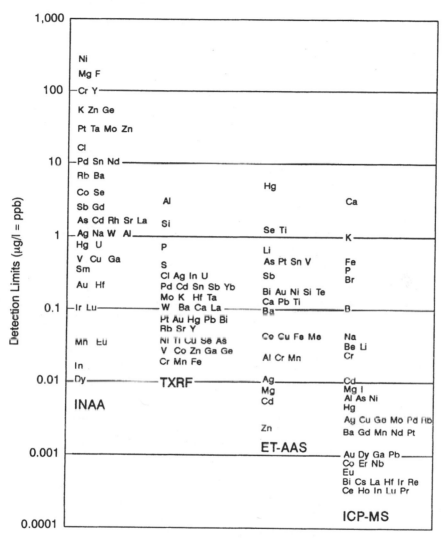

Figure 6-2. Relative detection limits of INAA, TXRF, ET-AAS, and ICP-MS, applied to trace analysis of aqueous solutions or their residues after evaporation. A 50 μL specimen was used for TXRF and ET-AAS; 3 mL were needed for INAA and ICP-MS. The individual values should be considered approximate at best. (Data from Tölg and Klockenkämper [10], Slavin [14], and Ehmann and Vance [15].)

must be diluted only for ICP-MS but can be directly analyzed by TXRF. No such restrictions are necessary for INAA. The total sample volume applied for analysis is consumed by ICP-MS, whereas TXRF and INAA are nonconsumptive.

Table 6-3. Comparison of Important Analytical Features of the Three Competitive Methods: ICP-MS, TXRF, and INAA

Analytical Features	ICP-MS	TXRF	INAA
Samples:			
Volume or mass	2–5 mL	5–50 μL	10–200 mg
Preparation of solids	Digestion or suspension	Digestion or suspension	None
Dissolvation portion	< 0.4%	< 1%	Any
Dilution of acids	1:100	None	None
Consumption	Yes	No	No
Detection:			
Detection limits	Excellent	Very good	Very good
Element limitations	H, C, N, O, F, P, S	$Z < 13$	$Z < 9$; Tl, Pb, Bi
Spectral interferences	Several	Few	Few
Isotope detection	Yes	No	No
Quantification:			
Calibration	Several external and internal standards	One internal standard	Some pure element foils
Matrix effects	Severe	None	None
Memory effects	Yes	No	No
Time consumption	< 3 min	< 20 min	20 min–30 days
Expenditure:			
Equipment	Ar-plasma + quadrupole MS	Special EDS	Nuclear reactor + γ-spectrometer
Capital costs	Medium	Medium	Very high
Running costs	High	Low	High
Maintenance	Frequently	Seldom	Seldom

TXRF and INAA permit detection at the low-pg level; ICP-MS has detection limits that can be lower by 1 or 2 orders of magnitude. Limitations exist for certain sets of elements, which are different for the three methods. While TXRF is limited in the detection of light elements with $Z < 13$, ICP-MS is hampered in the determination of some elements introduced by the air, e.g., N and O. For some other elements like H, F, P, and S, the ionization probability in the plasma is rather low. A majority of elements can also be determined by INAA, but again there are some exceptions. Light elements with $Z < 9$ and some heavy elements, e.g., Tl, Pb, and Bi, are not detectable without an additional device or a chemical procedure. Spectral interferences mostly occur for ICP-MS. Molecular ions build by the solvent or the carrier gas (e.g., argides) can lead to numerous overlaps. On the other hand, ICP-MS is capable of isotope analysis. A high degree of reliability in quantitative analysis can be achieved by isotope dilution.

For quantification, a highly elaborate calibration must be carried out in ICP-MS. Several external standards have to be used, and some internal standards may have to be added as well. In order to reduce the severe matrix effects, the external standards have to be adapted to the sample solution with respect to concentration and even acidity. Matrix effects can also be corrected for by standard addition and isotope dilution. But even then, memory effects and the already mentioned spectral overlaps can restrict the accuracy of ICP-MS.

In contrast to ICP-MS, the quantification for TXRF is much easier and more reliable. An internal standardization with a single element is possible. Neither matrix nor memory effects obstruct TXRF analysis, and the same is true for INAA. On the other hand, ICP-MS is a very fast method whereas INAA is often time consumptive because of the long irradiation and decay times in addition to the measuring time.

Regarding costs, INAA requires a high neutron flux for high detection power. It is only available at a nuclear reactor that exceeds the laboratory standard. ICP-MS and TXRF only need customary equipment of a medium size and price. But the capital expenditure for ICP-MS can double if a focusing mass spectrometer is used instead of a simple quadrupole MS. Such an instrument with high spectral resolution is recommended in order to avoid troublesome interferences. TXRF, in contrast, has moderate capital expenditures and also requires quite low running costs and little maintenance. By contrast, the Ar consumption of ICP-MS is a serious item (some $10,000 per year). The nebulizer, skimmer, pumps, and gas supply need frequent maintenance. For TXRF, liquid nitrogen (about $500 per year) and flat carriers are needed. But the carriers may be reused after cleaning.

In the field of surface and near-surface layer analyses, TXRF can compete quite well with conventional methods like Rutherford backscattering (RBS) or secondary ion mass spectrometry (SIMS). Only a few comparisons have as yet been published [17]. TXRF is nondestructive, needs no vacuum, and has no charging problems. The instrumentation is comparatively simple. It is a unique feature of TXRF that the density of near-surface layers can be determined in addition to their composition and thickness. The depth resolution amounts to several nm. Applications are mainly restricted to thin layers on optically flat samples such as wafers. But efforts have been made to analyze samples with a certain degree of roughness (up to some 10 or 100 nm). TXRF has even surpassed SIMS in the contamination control of wafers. The variant of VPD–TXRF shows the highest sensitivity (about 10^7 atoms/cm^2), particularly suitable to meet the high-purity demands in this field.

6.2. TRENDS AND FUTURE PROSPECTS

The efficiency of TXRF may well be increased by various recent instrumental and methodological developments. For example, several efforts have been made to extend the range of low-Z elements, to further reduce the detection limits for trace element analyses, to facilitate a lateral resolution for microdistribution analyses, to improve the vertical resolution for depth-distribution analyses, or even to obtain data on crystal structure (structure analysis) or chemical state (species analysis).

For the last two purposes, studies on the combination of grazing incidence with other spectrometrical methods, e.g., X-ray diffraction, X-ray absorption, and X-ray photoelectron emission, have been published [18–26]. Also these methods benefit from a reduced background at grazing incidence and lead to an improved structure or species analysis. (A detailed consideration of this subject, however, goes beyond the scope of this book.)

An ideal supplement to TXRF is the method of X-ray reflectometry (XRR) [9, 27–29]. Both methods allow the absolute determination of density, thickness, and even roughness of layered materials. However, the element composition of the layers can only be determined by TXRF, whereas the layer thickness can be measured more precisely by XRR. If the results of the individual fits are not in accord but ambiguous at first, both methods can work in tandem to find the true model of a layered material. A combined piece of equipment that allows for both TXRF and XRR measurements can easily be assembled, although it has not yet been made commercially available.

6.2.1. Instrumental Modifications

Modifications or extensions of the customary instrumentation commonly aim at improving the sensitivity. A special objective is to achieve an effective detection of the light elements with $Z < 14$. Instead of conventional X-ray tubes, synchrotron radiation may be used for excitation. But the respective large-scale storage rings for electrons or positrons necessary for this purpose are restricted to a number of about 20 worldwide.

6.2.1.1. Detection of Light Elements

Several obstacles hamper the detection of light elements especially with $Z < 14$. These elements, e.g., C, N, O, F, Na, Mg, and Al, have K-peaks in the low-energy region between 0.25 and 1.5 keV. The problems here are related to fluorescence excitation, energy-dispersive detection, and quantitative evaluation.

i. The fluorescence signals for light elements are 3–4 orders of magnitude smaller compared with medium-Z elements. The low sensitivity determined by equation (2-37), is caused by a reduced fluorescence yield ω, in accord with equation (1-5) and Figure 1-6, and a decreased photoelectric mass-absorption coefficient (τ/ρ), in accord with equation (1-14). This drawback is an inherent restriction.

ii. The spectral background in the low-energy region is high due to the scattered radiation of the X-ray tube. Its continuum is especially responsible, with a natural steep rise at low energies, in accord with equation (1-7) and Figure 1-7.

iii. The detector efficiency is bad in the low-energy region, as demonstrated in Figure 3-12. The spectral resolution is extensively limited by the electronic noise, in accord with equation (3-13).

iv. The absorption of the primary X-rays and the fluorescence X-rays with low photon energies is comparatively high. The absorption by ambient air and by the windows of the X-ray tube and detector leads to a reduced intensity.

v. The absorption of low-energy photons within the sample matrix can lead to matrix effects that give rise to difficulties in quantification.

Most of these problems are inherent to energy-dispersive XRF and cannot be alleviated. The others have been tackled in several attempts [30–35], especially by Streli et al. Specifically designed instrumentation is absolutely essential. Excitation by the K-radiation of a Cr- or a Sc-tube is recommended, and the X-ray tube window should be as thin as possible (8 μm Kapton foil) or even be absent. Instead of a low-pass filter, a multilayer should be used as monochromator. A HPGe detector is preferable to a Si(Li) detector, possibly equipped with an ultrathin diamond-like window of 0.4 μm. In addition, a special low-noise amplifier and preamplifier should be integrated. A vacuum chamber is of course mandatory, but a pressure of 0.1–1 hPa is sufficient.

All these efforts have produced absolute detection limits in the low-ng region: 0.4 ng Mg, 0.5 ng Na, 3.4 ng F, 10 ng O, and 13 ng C as reported by Streli et al. [33]). Relative detection limits for concentrations were estimated from a spectrum of NIST 1643c "Water" [35]. They are at or below the 0.1 μg/mL level for a counting time of 1000 s (0.03 μg/mL Al, 0.4 μg/mL Mg, and 0.5 μg/mL Na).

In order to reduce the detection limits further, a synchrotron beam was used for excitation (see also Section 6.2.1.2, below). Because of the high spectral brightness, detection limits of only a few pg were reached (0.5 pg Al, 2 pg Mg, and 9 pg F), according to Streli et al. [35], and even lower values (0.2 pg Mg) were reported recently by the same laboratory [36].

Quantification will suffer from matrix effects caused by unpredictable absorption in the low-energy range [37]. The conditions mentioned in Section 4.3.3 and the consequent limitations are more severe now. For micro- and trace analyses, the standard-addition technique might be applicable to this problem. For surface and thin-layer analyses, the fundamental parameter approach considered in Section 4.4.3 should be applied and angle-dependent intensity profiles must be recorded [38].

To improve the spectral resolution within the low-energy range it has been proposed that the energy-dispersive detector be replaced by a wavelength-dispersive device [37]. This improvement of resolution, however, would be bought at the expense of a loss in sensitivity, so that the detection power would decline.

6.2.1.2. Excitation by a Synchrotron

The lowest detection limits can be achieved by using a synchrotron beam for excitation. Three advantages of using this ideal excitation source should be noted:

i. The high spectral brightness of a synchrotron beam results in a primary X-ray intensity that is increased by 3–5 orders of magnitude compared to that of conventional X-ray tubes.

ii. Since synchrotron radiation covers a broad energy range of the X-ray spectrum, it is highly suitable for energy tuning. A multilayer crystal is preferably used for the selection of a particular energy band (see Section 3.3.2), which can easily be adjusted to an optimum excitation energy of the elements sought.

iii. The spectral background already reduced by total reflection is lowered further because of the polarization. The synchrotron beam is linearly polarized at 100% in the horizontal plane of the storage ring. To make use of this effect, the arrangement shown in Figure 6-3 should be set up. The detector should be placed in the horizontal plane with its axis normal to the synchrotron beam. The sample carrier can be positioned horizontally or vertically [39]. With this configuration, the background can reach its ultimate lowest limit determined by the bremsstrahlung of emitted photoelectrons [40].

High primary intensity, adjusted excitation energy, and the lowest possible background have resulted in detection limits lower by as much as 2 orders of magnitude compared to conventional excitation. The fg level has been reached, with 20 fg Sr, 30 fg Ni, and 150 fg Cd [41] and 200 fg for Mg [36]. In wafer analysis, detection limits between 3×10^8 and 1×10^9 atoms/cm^2 have been obtained for metallic contaminants [42, 43]. Furthermore, basic principles have been described and practical examples given for the depth analysis of impurities and for the characterization of multilayered structures [44–47]. However, much more widespread use of synchrotron radiation in the future

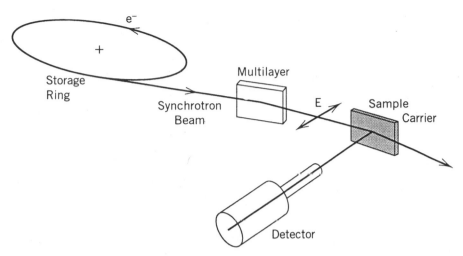

Figure 6-3. Arrangement for TXRF with excitation by a synchrotron beam. This beam is linearly polarized, i.e., the electric-field strength **E** lies in the plane of the electron (or positron) orbit. For a low spectral background, the detector axis must be placed in this plane and directed perpendicular to the beam itself.

will require easier access to such storage rings and the installation of suitable experimental stations.

6.2.2. Ablation and Deposition Techniques

Trace elements from an aqueous solution can be deposited on a TXRF carrier by an electrochemical reaction. The carrier should serve as a cathode in a dc cell with a continuous flow of the electrolyte. A conductive material such as glassy carbon is suitable as both the cathode and the TXRF carrier. Such electrochemical enrichment is still in its infancy but very promising [48–50].

The opposite process is the anodic decomposition of a metallic surface, also called electropolishing. It may be combined with a subsequent TXRF analysis of the eletrolyte. A high current density and a low temperature are beneficial to the fine etching of thin metallic layers. But the depth resolution seems to be restricted to a layer thickness of about 0.1 μm [51].

Native oxide layers on silicon surfaces can be decomposed by hydrofluoric acid, as noted in Section 5.4.4.2. The aqueous converted solution can be analyzed by TXRF. Continually repeated oxidation, ablation, and analysis of the uppermost layer might give a depth resolution of 1–3 nm.

In addition to such chemical or electrochemical methods used for depth profiling, physical methods have been proposed. An initial suggestion has

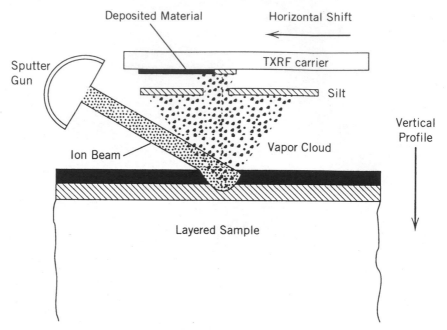

Figure 6-4. Sputter-etching device for vertical depth profiling. The layered sample is etched by an ion beam, and the sampled material is deposited on a carrier shifted horizontally behind a slit. Then, the carrier with the deposited thin layer is placed in a TXRF instrument and analyzed step by step. (After Schwenke et al. [37].)

come from Schwenke et al. [37], and a patent has been applied for [52]. Figure 6-4 depicts the procedure schematically. The layered sample is first etched or eroded by an ion beam of a sputter device, usually by an Ar^+-beam. This process is commonly used in combination with methods of surface or thin-layer analysis, e.g., AES, XPS, or SIMS. The sputtered material emitted as an atomic vapor partly passes a shielding slit positioned above the sample. The vapor is deposited as a thin layer on a substrate suitable as a TXRF carrier.

The substrate is horizontally shifted during the sputter process, and thereby the vertical concentration profile is transformed into a horizontal one. After-ward, the thin layer with its lateral distribution of atoms is subjected to TXRF analysis. The substrate is placed under the aperture (< 5 mm) of the energy-dispersive detector and again horizontally shifted, step by step. A lateral scan is recorded, which after calibration gives the depth profile as desired.

This method is rather promising since it is applicable to multilayer struc-tures with a thickness of some 10 nm up to several hundred nm. The samples need not to be flat and even, but may have rough or wavy surfaces. The depth resolution might be on the order of 5 nm. An additional variant is offered by use of a collimated ion beam instead of a broad ion beam or even by a focused laser beam [53]. Material can be sputtered from a small spot, deposited on a carrier, and analyzed by TXRF. In this way, a microdistribution analysis becomes possible. By successive ablations from neighboring spots, a line scan can be recorded. The lateral resolution might be about $5\,\mu$m. Above all, the instrumental device is rather simple and a high vacuum is not necessary.

A second approach to depth-profile analysis is also based on sputtering. The subsequent analysis, however, is not applied to the sample material removed but to that which remains. The method was first suggested by Knoth et al. [54] and promptly applied by Wiener et al. [55] and Frank et al. [56]. First, the samples are sputter-etched by a wide ion source of 4–5 cm diameter. A clean argon gas is used, adjusted to a pressure of 5×10^{-2} Pa, thus allowing a high ion flux up to $1\,mA/cm^2$ with a relatively low energy of 500 eV. As illustrated in Figure 6-5, the wide ion beam is directed perpendicular to the sample.

During sputtering, the sample is more and more shielded by a tantalum shutter which is steadily moved in front of the sample [55]. In a modified technique, the sample is moved along a straight line behind a shield with an aperture of about $1\,cm^2$ [56]. Both techniques etch the sample to a bevel with an inclination angle of $< 0.0001°$. The vertical depth scale is thereby transformed to a horizontal length scale at a magnification of about 10^6; that is,

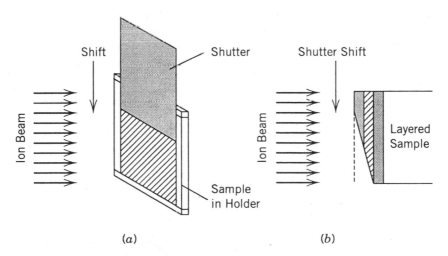

Figure 6-5. Ion beam sputtering for a bevel cut of a layered sample: (a) during the sputter process, a tantalum shutter is steadily moved down; (b) the sample is etched to a bevel shape and then subjected to a lateral scan with TXRF. (After Wiener et al. [55].)

layers of 10 nm thickness might give stripes of 1 cm width. Layers of such a thickness are sputtered within 1–10 min.

After the sputter process, the beveled samples are moved in steps below a TXRF detector with a diaphragm of 0.2–1 mm width. The angle of incidence is chosen to be well below the critical angle of the relevant elements (about 0.1–0.3°). The intensity for the individual elements is recorded against the lateral position, producing a laterally resolved line scan. To derive a concentration vs. depth profile, both axes have to be calibrated. The length axis or abscissa is calibrated by a determination of the different thicknesses of the individual layers. This can easily be done by an additional measurement with a profilometer [56]. Or it may be done by an additional experiment with TXRF. This time, however, the incident angle should be set far above the critical angle [55]. The intensity axis or ordinate can be calibrated by the algorithm described in Section 4.4.3. The fluorescence intensity is then calculated for the remaining layers, dependent on the sputter depth.

The foregoing method is not only applicable to flat layered samples with sharp interfaces as a prerequisite for nondestructive depth-profiling by TXRF; it is also applicable to layered samples with diffuse interfaces. A depth resolution of 2–3 nm may be reached, and a sampling depth of up to several hundred nm.

6.2.3. Grazing Exit Arrangements

Recent developments feature TXRF analysis at the grazing exit either *instead of* the grazing incidence [57–59] or *in addition to* the grazing incidence [59–62]. These methods are highlighted in Figure 6-6. The conventional method is illustrated in part (a) of the figure. The primary beam is incoming at the grazing incidence and totally reflected at the sample carrier. The fluorescence radiation is detected perpendicular to the carrier. This (0°,90°) configuration is changed into a (90°,0°) configuration in part (b). The angle of incidence and the takeoff angle are interchanged. In part (c), the primary beam is directed toward the carrier at the grazing incidence and the fluorescence beam is taken off at the grazing exit. Both angles are about zero in this (0°,0°) arrangement, while the angle between both beams is 90°. The three modifications may be called GI-XRF (grazing incidence X-ray fluorescence), GE-XRF (grazing exit XRF) and GIE-XRF (grazing incidence and exit XRF). Any further distinction between a fixed and a variable grazing angle seems to be unnecessary.

All three techniques can involve the effect of total reflection if the relevant angle is below the corresponding critical angle. This is shown for GE experiments by Figure 6-7. The atoms excited to fluorescence by the primary beam may be placed above a substrate or within a layer on top of the substrate. In

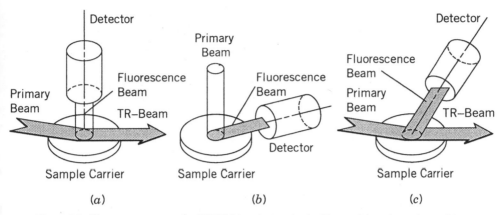

Figure 6-6. Three arrangements for TXRF: (a) at the grazing incidence of the primary beam; (b) at the grazing exit of the fluorescence beam; and (c) at the grazing incidence and exit of both beams. (TR = total-reflection.)

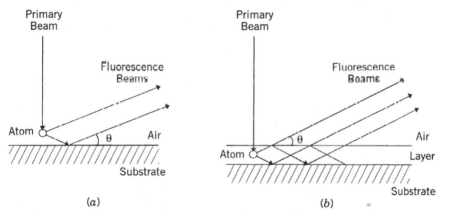

Figure 6-7. Beam paths for X-ray fluorescence at grazing exit. The atom excited by the primary beam is placed (a) above the substrate or (b) within a layer on top of the substrate. Different beams emitted from the atom may be reflected at the substrate and may interfere with each other.

any case, different coherent fluorescence beams may be parallel after reflection at the substrate and so interfere with each other. The interference pattern will be highly distinctive if total reflection occurs at or below the critical angle.

The analytical procedure for all three techniques is the same. Micro- or trace analyses are performed after a small sample amount is put onto a suitable glass carrier at a fixed angle. Surface and thin-layer analyses are carried out by tilting the flat sample and by recording angle-dependent intensity profiles. The following differences should be considered:

i. Under the GE condition the critical angle is determined by the fluorescence radiation, whereas under the GI condition it is determined by the primary radiation. According to equation (1-34), the critical angle for the GE is larger than that for the GI inasmuch as the respective photon energy is smaller.

ii. The information depth under the GE condition corresponds to the *emergence* depth of the fluorescence beam and is somewhat smaller than that under the GI condition, which corresponds to the *penetration* depth of the primary beam.

Back in 1983, Becker et al. performed GE experiments for surface analysis [63]. Thereafter, the GE technique was used to characterize submonolayers of adsorbates [64] and implantation profiles in wafers [65]. Angle-dependent profiles with an oscillation structure were observed for thin layers on a substrate [58–60]. The results obtained for GE-XRF were shown to agree quite well with those obtained by GI-XRF. For such a comparison, the reciprocity theorem of optics named for H.L.F. von Helmholtz (see Born and Wolf [66]) can be applied. This theorem furthermore simplifies the calculation of fluorescence intensities, which can be based on the recursive formalism (see Section 2.4). A straightforward calculation is also possible but needs an asymptotic analysis of the Maxwell equations in a stratified layer [67].

The new variants of GE- or GIE-XRF were mostly employed in nondestructive surface or thin-layer analysis using a synchrotron beam for excitation. A monochromatic excitation (such as is needed for GI-XRF) is not necessary. This is the first distinct advantage of the GE configuration. Moreover, a collimated or focused microbeam of the synchrotron radiation can be applied, offering a lateral resolution of a few μm. Concentration profiles along straight lines or area maps can be recorded by scanning the sample. This possibility is excluded for GI-XRF since the grazing incidence is an inherent obstacle to a high spatial resolution.

A further advantage of GE-XRF is the possibility of replacing the energy-dispersive detector by a wavelength-dispersive detector. The choice of a crystal spectrometer enables a more reliable detection of light elements because of far better resolution in the low-energy spectrum. Such a combination is suitable for GE-XRF but not for GI-XRF due to intensity limitations. On the other hand, absorption effects become more severe under GE conditions and may diminish the advantage. The large path length of the emerged beam with photons of lower energy can cause matrix effects and will be detrimental to the quantitative determination of elements.

In comparison to GI and GE instrumentation, GIE-XRF makes still higher experimental demands. Both the incidence and takeoff angles have to be controlled with an accuracy of 0.001°. Furthermore, the fluorescence intensity is minimal since both the incidence beam and the exit beam are strongly restricted in their divergence. On the other hand, GIE-XRF allows the variation of a second decisive angle. The incident angle can be varied at different fixed takeoff angles and vice versa. The various combinations allow cross-checking and may yield additional information [59].

REFERENCES

1. Prange, A. (1989). *Spectrochim. Acta* **44B**, 437.

2. Klockenkämper, R. (1991). *In Analytiker Taschenbuch* (H. Günzler, R. Borsdorf, W. Fresenius, W. Huber, H. Kelker, I. Lüderwald, G. Tölg, and H. Wisser, eds.), Vol. 10, p. 111. Springer, Berlin.

3. Klockenkämper, R., and von Bohlen, A. (1992). *J. Anal. At. Spectrom.* **7**, 273.

4. Prange, A., and Schwenke, H. (1992). *Adv. X-Ray Anal.* **35B**, 899.

5. Klockenkämper, R., Knoth, J., Prange, A., and Schwenke, H. (1992). *Anal. Chem.* **64**, 1115A.

6. Schwenke, H., and Knoth, J. (1993). "Total Reflection XRF," *In Handbook on X-Ray Spectrometry* (R. van Grieken and A. Markowicz, eds.) Practical Spectroscopy Series, Vol. 14, p. 453. Dekker, New York.

7. van den Hoogenhof, W.W., and de Boer, D.K.G. (1993). *Spectrochim. Acta* **48B**, 277.

8. Taniguchi, K. (1993). *Bunseki* 3, 168.

9. de Boer, D.K.G., Leenaers, A.J.G., and van den Hoogenhof, W.W. (1995). *X-Ray Spectrom.* **24**, 91.

10. Tölg, G., and Klockenkämper, R. (1993). *Spectrochim. Acta* **48B**, 111.

11. Michaelis, W. (1986). *Fresenius' Z. Anal. Chem.* **324**, 662.

12. Krivan, V., Schneider, G., Baumann, H., and Reus, U. (1994). *Fresenius' J. Anal. Chem.* **348**, 218.

13. Pepelnik, R., Prange, A., and Niedergesäss, R. (1994). *J. Anal. At. Spectrom.* **9**, 1071.

14. Slavin, W. (1992). *Spectrosc. Int.* **4**, 22.

15. Ehmann, W.D., and Vance, D.E. (1991). *Radiochemistry and Nuclear Methods of Analysis.* Wiley, New York.

16. Lieser, K.H., Flakowski, M., and Hoffmann, P. (1994). *Fresenius' J. Anal. Chem.* **350**, 135.

17. Penka, V., and Hub, W. (1989). *Spectrochim. Acta* **44B**, 483.

18. Marra, W.C., Eisenberger, P., and Cho, A.Y. (1979). *J. Appl. Phys.* **50**, 6927.

19. Segmüller, A. (1987). *Thin Solid Films* **154**, 33.

20. Feidenhans'l, R. (1989). *Surf. Sci. Rep.* **10**, 105.

21. Huang, T.C. (1992). *Adv. X-Ray Anal.* **35A**, 143.

22. Wulff, H., Klimke, J., and Quade, A. (1995). *GIT Fachz. Lab.* **39**, 1063.

23. Horiuchi, T., and Matsushige, K. (1993). *Spectrochim. Acta* **48B**, 137.

24. Kawai, J., Takami, M., Fujinami, M., Hashigushi, Y., Hayakawa, S., and Gohshi, Y. (1992). *Spectrochim. Acta* **47B**, 983.

25. Kawai, J., Adachi, H., Hayakawa, S., Zheng, Z., Kobayashi, K., Gohshi, Y., Maeda, K., and Kitayima, Y. (1994). *Spectrochim Acta* **49B**, 739.

26. Kawai, J., Hayakawa, S., Kitajima, Y., and Gohshi, Y. (1995). *Adv. X-Ray Chem. Anal. Jpn.* **26s**, 97.

27. Huang, T.C., and Parrish, W. (1992). *Adv. X-Ray Anal.* **35A**, 137.

28. Lengeler, B. (1992). *Adv. X-Ray Anal.* **35A**, 127.

29. Hüppauf, M. (1993). Ph.D. thesis, RWTH Aachen.

30. Streli, C., Aiginger, H., and Wobrauschek, P. (1989). *Spectrochim. Acta* **44B**, 491.

31. Streli, C., Wobrauschek, P., and Aiginger, H. (1992). *Adv. X-Ray Anal.* **35B**, 947.

32. Hein, M., Hoffmann, P., Lieser, K.H., and Ortner, H.M. (1992). *Fresenius' J. Anal. Chem.* **343**, 760.

33. Streli, C., Aiginger, H., and Wobrauschek, P. (1993). *Spectrochim. Acta* **48B**, 163.

34. Streli, C., Aiginger, H., and Wobrauschek, P. (1993). *Nucl. Instrum. Methods* **A334**, 425.

35. Streli, C., Wobrauschek, P., Ladisich, W., Rieder, R., and Aiginger, H. (1995). *X-Ray Spectrom.* **24**, 137.

36. Streli, C., Wobrauschek, P., Ladisich, W., Rieder, R., Aiginger, H., Ryon, R.W., and Pianetta, P. (1994). *Nucl. Instrum. Methods* **A345**, 399.

37. Schwenke, H., Bormann, R., Knoth, J., and Prange, A. (1993). *Spectrochim. Acta* **48B**, 293.

38. Streli, C., Wobrauschek, P., Randolf, G., Rieder, R., Ladisich, W., and Aiginger, H. (1995). *Adv. X-Ray Chem. Anal. Jpn.* **26s**, 63.

39. Wobrauschek, P., Kregsamer, P., Ladisich, W., Rieder, R., and Streli, C. (1993). *Spectrochim. Acta* **48B**, 143.

40. Takaura, N., Brennan, S., Pianetta, P., Laderman, S.S., Fischer-Colbrie, A., Kortright, J.B., Wherry, D.C., Miyazaki, K., and Shimazaki, A. (1995). *Adv. X-Ray Chem. Anal. Jpn.* **26s**, 113.

41. Rieder, R., Wobrauschek, P., Ladisich, W., Streli, C., Aiginger, H., Garbe, S., Gaul, G., Knöchel, A., and Lechtenberg, F. (1995). *Nucl. Instrum. Methods* **A355**, 648.

42. Laderman, S.S., Fischer-Colbrie, A., Shimazaki, A., Miyazaki, K., Brennan, S., Takaura, N., Pianetta, P., and Kortright, J.B. (1995). *Adv. X-Ray Chem. Anal. Jpn.* **26s**, 91.

43. Liu, K.Y., Kojima, S., Kawado, S., and Iida, A. (1995). *Adv. X-Ray Chem. Anal. Jpn.* **26s**, 107.

44. Iida, A., Yoshinaga, A., Sakurai, K., and Gohshi, Y. (1986). *Anal. Chem.* **58**, 394.

45. Matsushita, T., Iida, A., Ishikawa, T., Nakagiri, T., and Sakai, K. (1986). *Nucl. Instrum Methods* **A246**, 751.

46. Iida, A. (1991). *Adv. X-Ray Anal.* **34**, 23.

47. Iida, A. (1992). *Adv. X-Ray Anal.* **35B**, 795.

48. Kollotzek, D. (1980). *Diploma thesis*, University of Stuttgart.

49. Eller, R., Alt, F., Tölg, G., and Tobschall, H.J. (1989). *Fresenius' Z. Anal. Chem.* **334**, 723.

50. Fan, Q., and Gohshi, Y. (1993). *Appl. Spectrosc.* **47**, 1742.

51. Krämer, K. (1982). Ph.D. thesis, University of Stuttgart.

52. Bormann, R., and Schwenke, H. (1992). Offenlegungsschrift DE 4,028,044 A1, Int. Cl. G 01 N23/203, Deutsches Patentamt.

53. Bredendiek-Kämper, S., von Bohlen, A., Klockenkämper, R., Quentmeier, A., and Klockow, D. (1996). To be published in *J. Anal. At. Spectrom.*

54. Knoth, J., Bormann, R., Gutschke, R., Michaelsen, C., and Schwenke, H. (1993). *Spectrochim. Acta* **48B**, 285.

55. Wiener, G., Michaelsen, C., Knoth, J., Schwenke, H., and Bormann, R. (1995). *Rev. Sci. Instrum.* **66**, 20.

56. Frank, W., Thomas, H.-J., and Schindler, A. (1995). *Spectrochim. Acta* **50B**, 265.

57. Sasaki, Y., and Hirokawa, K. (1990). *Appl. Phys. A* **50**, 397.

58. Noma, T., and Iida, A. (1994). *Rev. Sci. Instrum.* **65**, 837.

59. Noma, T., Iida, A., and Sakurai, K. (1993). *Phys. Rev. B* **48**, 17524.

60. Tsuji, K., and Hirokawa, K. (1994). *J. Appl. Phys.* **A75**, 7189.

61. Tsuji, K., Sato, S., and Hirokawa, K. (1995). *Adv. X-Ray Chem. Anal. Jpn.* **26s**, 151.

62. Sasaki, Y. (1995). *Adv. X-Ray Chem. Anal. Jpn.* **26s**, 193.

63. Becker, S., Golovchenko, J.A., and Patel, J.R. (1983). *Phys. Rev. Lett.* **50**, 153.

64. Hasegawa, S., Ino, S., Yamamoto, Y., and Daimon, H. (1985). *Jpn. J. Appl. Phys.* **24**, L387.

65. Sasaki, Y.C., and Hirokawa, K. (1991). *Appl. Phys. Lett.* **58**, 1384.

66. Born, M., and Wolf, E. (1980). *Principles of Optics*. Pergamon Press, New York, 6th ed., reprinted 1993.

67. de Bokx, P.K., and Urbach, H.P. (1995). *Adv. X-Ray Chem. Anal. Jpn.* **26s**, 199.

INDEX